LONDON MATHEMATICAL SOCIETY L

Managing Editor: Professor J.W.S. Cassels, Depa ... istics,
University of Cambridge, 16 Mill Lane, Cambridg

The books in the series listed below are available fr
from Cambridge University Press.

4 Algebraic topology, J.F. ADAMS
17 Differential germs and catastrophes, Th. BROCKER & L. LANDER
27 Skew field constructions, P.M. COHN
34 Representation theory of Lie groups, M.F. ATIYAH *et al*
36 Homological group theory, C.T.C. WALL (ed)
39 Affine sets and affine groups, D.G. NORTHCOTT
40 Introduction to H_p spaces, P.J. KOOSIS
42 Topics in the theory of group presentations, D.L. JOHNSON
43 Graphs, codes and designs, P.J. CAMERON & J.H. VAN LINT
45 Recursion theory: its generalisations and applications, F.R. DRAKE & S.S. WAINER (eds)
46 p-adic analysis: a short course on recent work, N. KOBLITZ
49 Finite geometries and designs, P. CAMERON, J.W.P. HIRSCHFELD & D.R. HUGHES (eds)
50 Commutator calculus and groups of homotopy classes, H.J. BAUES
51 Synthetic differential geometry, A. KOCK
54 Markov processes and related problems of analysis, E.B. DYNKIN
57 Techniques of geometric topology, R.A. FENN
58 Singularities of smooth functions and maps, J.A. MARTINET
59 Applicable differential geometry, M. CRAMPIN & F.A.E. PIRANI
60 Integrable systems, S.P. NOVIKOV *et al*
62 Economics for mathematicians, J.W.S. CASSELS
65 Several complex variables and complex manifolds I, M.J. FIELD
66 Several complex variables and complex manifolds II, M.J. FIELD
68 Complex algebraic surfaces, A. BEAUVILLE
69 Representation theory, I.M. GELFAND *et al*
74 Symmetric designs: an algebraic approach, E.S. LANDER
76 Spectral theory of linear differential operators and comparison algebras, H.O. CORDES
77 Isolated singular points on complete intersections, E.J.N. LOOIJENGA
78 A primer on Riemann surfaces, A.F. BEARDON
79 Probability, statistics and analysis, J.F.C. KINGMAN & G.E.H. REUTER (eds)
80 Introduction to the representation theory of compact and locally compact groups, A. ROBERT
81 Skew fields, P.K. DRAXL
82 Surveys in combinatorics, E.K. LLOYD (ed)
83 Homogeneous structures on Riemannian manifolds, F. TRICERRI & L. VANHECKE
84 Finite group algebras and their modules, P. LANDROCK
85 Solitons, P.G. DRAZIN
86 Topological topics, I.M. JAMES (ed)
87 Surveys in set theory, A.R.D. MATHIAS (ed)
88 FPF ring theory, C. FAITH & S. PAGE
89 An F-space sampler, N.J. KALTON, N.T. PECK & J.W. ROBERTS
90 Polytopes and symmetry, S.A. ROBERTSON
91 Classgroups of group rings, M.J. TAYLOR
92 Representation of rings over skew fields, A.H. SCHOFIELD
93 Aspects of topology, I.M. JAMES & E.H. KRONHEIMER (eds)
94 Representations of general linear groups, G.D. JAMES
95 Low-dimensional topology 1982, R.A. FENN (ed)

96 Diophantine equations over function fields, R.C. MASON
97 Varieties of constructive mathematics, D.S. BRIDGES & F. RICHMAN
98 Localization in Noetherian rings, A.V. JATEGAONKAR
99 Methods of differential geometry in algebraic topology, M. KAROUBI & C. LERUSTE
100 Stopping time techniques for analysts and probabilists, L. EGGHE
101 Groups and geometry, ROGER C. LYNDON
103 Surveys in combinatorics 1985, I. ANDERSON (ed)
104 Elliptic structures on 3-manifolds, C.B. THOMAS
105 A local spectral theory for closed operators, I. ERDELYI & WANG SHENGWANG
106 Syzygies, E.G. EVANS & P. GRIFFITH
107 Compactification of Siegel moduli schemes, C-L. CHAI
108 Some topics in graph theory, H.P. YAP
109 Diophantine Analysis, J. LOXTON & A. VAN DER POORTEN (eds)
110 An introduction to surreal numbers, H. GONSHOR
111 Analytical and geometric aspects of hyperbolic space, D.B.A.EPSTEIN (ed)
112 Low-dimensional topology and Kleinian groups, D.B.A. EPSTEIN (ed)
113 Lectures on the asymptotic theory of ideals, D. REES
114 Lectures on Bochner-Riesz means, K.M. DAVIS & Y-C. CHANG
115 An introduction to independence for analysts, H.G. DALES & W.H. WOODIN
116 Representations of algebras, P.J. WEBB (ed)
117 Homotopy theory, E. REES & J.D.S. JONES (eds)
118 Skew linear groups, M. SHIRVANI & B. WEHRFRITZ
119 Triangulated categories in the representation theory of finite-dimensional algebras, D. HAPPEL
121 Proceedings of *Groups - St Andrews 1985*, E. ROBERTSON & C. CAMPBELL (eds)
122 Non-classical continuum mechanics, R.J. KNOPS & A.A. LACEY (eds)
123 Surveys in combinatorics 1987, C. WHITEHEAD (ed)
124 Lie groupoids and Lie algebroids in differential geometry, K. MACKENZIE
125 Commutator theory for congruence modular varieties, R. FREESE & R. MCKENZIE
127 New directions in dynamical systems, T.J. BEDFORD & J.W. SWIFT (eds)
128 Descriptive set theory and the structure of sets of uniqueness, A.S. KECHRIS & A. LOUVEAU
130 Model theory and modules, M. PREST
131 Algebraic, extremal & metric combinatorics, M-M. DEZA, P. FRANKL & I.G. ROSENBERG (eds)
132 Whitehead groups of finite groups, ROBERT OLIVER

London Mathematical Society Lecture Note Series. 113

Lectures on the Asymptotic Theory of Ideals

D. Rees, F.R.S.
Emeritus Professor of Mathematics,
University of Exeter

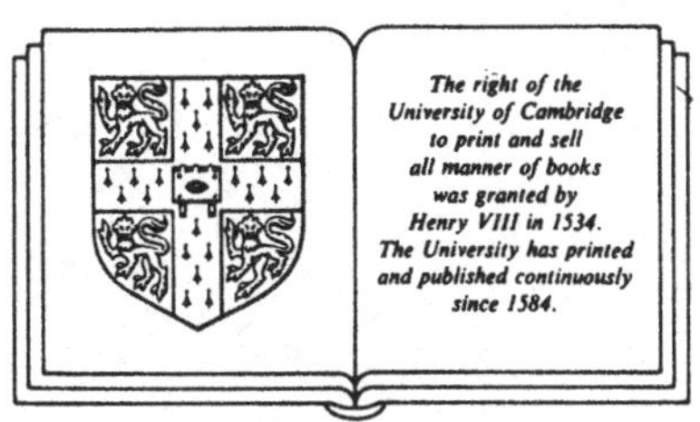
The right of the
University of Cambridge
to print and sell
all manner of books
was granted by
Henry VIII in 1534.
The University has printed
and published continuously
since 1584.

CAMBRIDGE UNIVERSITY PRESS
Cambridge

Published by the Press Syndicate of the University of Cambridge
The Pitt Building, Trumpington Street, Cambridge CB2 1RP
32 East 57th Street, New York, NY 10022, USA
10, Stamford Road, Oakleigh, Melbourne 3166, Australia

First published 1988

Library of Congress cataloging in publication data
Rees, D., 1918 -
Lectures on the asymptotic theory of ideals.
(London Mathematical Society lecture note series: 113)
Bibliography: p. Includes Indexes
1. Ideals (algebra) - Asymptotic theory.
I. Title II. Series
QA247.R39 1988
512'.4 87 – 35804

British Library cataloguing in publication data available

ISBN 0 521 31127 6

Transferred to digital printing 2001

TO JOAN

"Many daughters have done virtuously,
but thou excellest them all."
Proverbs, Chapter XXXI, v. 29

TABLE OF CONTENTS

PREFACE

The greater part of these notes were presented in the form of lectures during a visit to Nagoya University during the Winter of 1982-3. This visit was made possible by the support of the Japanese Ministry of Education and the British Council, and to both these bodies I would like to express my gratitude. My warmest thanks go to Professor Hideyuki Matsumura and his wife for their hospitality and the friendship they showed throughout my stay.

Coming closer to home, my thanks are also due to George Duller, Peter Vamos and the other members of the Mathematics Department of the University of Exeter for allowing me the use of a word processor. My thanks again to George Duller, and also to Tony Stratton, for the continual help and advice they have given me.

I also owe a debt of gratitude to the Cambridge University Press, particularly David Tranah, for advice throughout the preparation of the manuscript.

My thanks are also due to Professor Rodney Sharp, whose careful reading of the manuscript brought to light a great number of obscurities and errors which, I hope, have now been removed.

Finally, and not least, I thank my wife for her patience and her extraordinary ability as a proof-reader, and for a great deal more besides.

November 1987.

INTRODUCTION

In 1982, I was invited to give a course of 11 two-hour lectures in the University of Nagoya on some branch of Commutative Algebra. The topic I chose was the asymptotic theory of ideals and the lectures were duly given between December 1982 and March 1983. The notes below are an extensive revision of the notes given to the audience at the lectures and, with certain exceptions, the chapter headings below correspond to the titles of the individual lectures. The exceptions referred to are the following. First, the notes of the third lecture have been considerably expanded so as to incorporate a proof of the Mori-Nagata Theorem, based on the beautiful theorem of Matijevic, and the original topic of the third lecture, the Valuation Theorem, is dealt with in the fourth lecture. The second change is more considerable. The last three lectures of the course dealt with Teissier's theory of mixed multiplicities as given in Teissier[1973] and was based on the use of complete and joint reductions of a set of ideals. In the last lecture I applied these ideas to prove what I call the general degree formula. An account of the theory of complete and joint reductions has since appeared in Rees[1984], while, since the lectures were given, I have succeeded in proving a still more general degree formula using a quite different method. This method is the method of general elements of ideals and the last three chapters of these notes now deal with the theory of general elements finishing with a proof of the new version of the degree formula. These three chapters are therefore separate from the first nine and can almost be read independently. An earlier version of the material contained in these chapters appeared in Rees[1986].

We now consider the contents of the notes in more detail. The asymptotic theory of ideals originated with the paper Samuel[1952] which contained, in different language, most of the basic ideas lying behind what follows. We commence by considering some of these ideas, using the language of filtrations. By a *filtration* f on a commutative ring A we understand a function defined on A which takes real values, or the value ∞, satisfying the conditions which follow.

i) $f(1) \geq 0$, $f(0) = \infty$; ii) $f(x-y) \geq \mathrm{Min}(f(x),f(y))$; iii) $f(xy) \geq f(x) + f(y)$.

If f,g are filtrations, we consider them as equivalent if there is a finite constant K such that $|f(x)-g(x)| < K$ for all x. This is understood to mean that $g(x) = \infty$ if and only if $f(x) = \infty$. This implies that every filtration is equivalent to a filtration whose values are integers or ∞, and we often restrict attention to such filtrations. One of Samuel's ideas was to introduce a filtration $\mathbf{f}(x)$ defined as the limit as n tends to ∞ of $f(x^n)/n$. That this limit exists and is a filtration is proved in chapter 2. Samuel's main interest was in the case where A is noetherian and f is a filtration f_J associated with an ideal J and defined by $f_J(x) \geq n$ if and only if x belongs to J^n. Samuel conjectured that $\mathbf{f}_J(x)$, if finite, was rational. This was proved independently by M. Nagata[1956] and myself[1956b]. It is with the second of these proofs that we are now concerned. A stronger result was proved, referred to in these notes as the Valuation Theorem. This states that $\mathbf{f}_J(x)$ can be expressed in the form Min $v_i(x)/v_i(J)$, where v_i ranges over a finite set of valuations v_i, which take non-negative values on A, positive values on J, and the value ∞ on a minimal prime ideal $\mathbf{p}_i$ of A depending on v_i. The notation $v_i(J)$ denotes the minimal value of $v_i(x)$ on J.

The proof of the theorem depended on the introduction of an ancillary graded ring which we now describe and which in these notes is denoted by $G(f_J)$. This is the sub-ring of $A[t, t^{-1}]$ consisting of all finite sums $\Sigma c_r t^r$, summed, say, from -p to q, satisfying the condition that $f_J(c_r) \geq r$ for each r. Now write u for t^{-1} and G for $G(f_J)$. Then f_J is the restriction to A of the filtration f_{uG} on G and the proof of the Valuation Theorem is reduced to the special case where J is a principal ideal generated by a non-zero divisor. Since G is noetherian, this point being crucial, this special case is a fairly easy deduction from the Mori-Nagata Theorem. In fact the proof of [1956b], obtained in 1955, did not use the Mori-Nagata Theorem, since Nagata's proof of the general case in Nagata[1955] was not then available to me.

The definition of $G(f_J)$ can obviously be adapted to define a graded ring G(f) for any filtration f and the proof indicated above can also be adapted to prove a Valuation Theorem for f, providing that G(f) is noetherian. This leads to the introduction of a class of filtrations, noether filtrations, defined as those for which G(f) is noetherian, with the additional restriction that they take integer values together with ∞. Again a result of Samuel plays a crucial part, his characterisation of graded noetherian rings G as those for which the sub-ring G_0 of elements of degree zero is noetherian and which, in addition, are finitely generated over this sub-ring.

I first met this in Samuel[1953], and it is referred to below as Samuel's Theorem. This theorem enables us to describe noether filtrations in some detail, and this is done in chapter 2. We will merely give one consequence, which appears in chapter 6 as Lemma 6.11, to the effect that, if f is a noether filtration which takes only non-negative values, then it is equivalent to a filtration $w.f_J$ for some ideal J, and some positive integer w. This indicates the key role played by the filtrations f_J.

Now, at last, we consider the individual chapters, restricting ourselves to the first five chapters for the present. Chapter 1 collects together some general results on graded noetherian rings, based for the most part on Samuel's Theorem, but including an account of the theory of Hilbert Functions using the Koszul Complex. Chapter 2 is concerned with elementary results on filtrations, particularly noether filtrations, and we will pick out some of these. First, there is a uniqueness theorem for the representation of $\mathfrak{f}(x)$ in the form $\operatorname{Min} v_i(x)/e_i$, where v_i ranges over a finite set of valuations and the numbers e_i are real. Note that the existence of the representation does not appear until chapter 4. Next, in this chapter we associate with a noether filtration f another filtration f* which is integer-valued and closely associated with $\mathfrak{f}(x)$. This is the integral closure of f. It is defined by $f^*(x) \geq n$ if x satisfies an equation

$$x^r + c_1x^{r-1} + \ldots + c_r = 0$$

with $f(c_i) \geq ni$ for each i. f* is related to f, $\mathfrak{f}$ by inequalities

$$f(x) \leq f^*(x) \leq \mathfrak{f}(x) \leq f^*(x) + 1.$$

In fact $f^*(x)$ is the integral part of $\mathfrak{f}(x)$, but this is not proved until chapter 4. Finally, if f and g are two noether filtrations taking only non-negative values, then they are equivalent if and only if $f^*(x) = g^*(x)$ for all x.

Now we come to chapters 3 to 5. The first of these contains a proof of the theorem of Matijevic (Matijevic[1976]) and uses it to prove the Mori-Nagata Theorem (Mori[1952] for local domains and Nagata[1955] for general noetherian domains). The proof given here draws heavily on the papers of Querre[1979] and Kiyek[1981]. Chapter 4 is devoted to a proof of the Valuation Theorem for any noether filtration f. It takes the form

$$\mathfrak{f}(x) = \mathrm{Min}\, v_i(x)/v_i(f)$$

the minimum being taken over a finite set of valuations v_i. This theorem is proved for noether filtrations which may take negative values on A. In this case the definition of $v_i(f)$ is somewhat complicated and we restrict attention here to the case where f(x) takes only non-negative values. First we note that the valuations v_i take values which are non-negative integers or ∞ on A and positive values on the radical of f, this being defined as the set of elements for which $f(x^n) > 0$ for some n. If v_i is proper, that is, takes values other than $0,\infty$, then $v_i(f)$ is the minimum of $v_i(x)/f(x)$ taken over those x for which f(x) is neither 0 or ∞. If v_i is degenerate $v_i(f)$ can be taken to be 1.

Now we come to chapter 5. In this and the later chapters, f is restricted to take non-negative values. Chapter 5 has as its objective the Strong Valuation Theorem. The aim of this theorem is the determination of conditions under which $f^*(x)$ is a noether filtration. It states that this is true for all noether filtrations on A if and only if, for every maximal ideal $\mathfrak{m}$ of A, the local ring $A_{\mathfrak{m}}$ is analytically unramified, that is, the completion $(A_{\mathfrak{m}})^\wedge$ of $A_{\mathfrak{m}}$ has no nilpotent elements. This requires a great deal of the theory of completions, and hence chapter 5 contains a brief account of the relevant material, given without proofs.

In chapters 6 and 7, the problem considered is that of determining which proper valuations v are associated with some noether filtration f on A via the Valuation Theorem. Since f can be taken to be of the form f_J by Lemma 6.11, these are termed ideal valuations. The two chapters give different characterisations of ideal valuations. In chapter 6, the general case is reduced to the case where A is a local domain $(Q,\mathfrak{m},k,d)$ and v has radical $\mathfrak{m}$. Here $\mathfrak{m}$ is the maximal ideal of Q, $k = Q/\mathfrak{m}$, and d is the Krull dimension of Q. This implies that v has a unique extension $v^\wedge$ to the completion $Q^\wedge$ of Q. Then for v to be an ideal valuation it is necessary and sufficient that this extension $v^\wedge$ takes the value ∞ on a minimal prime ideal $\mathfrak{p}$ of $Q^\wedge$, and that the residue field K_v of v has transcendence degree $\dim(Q^\wedge/\mathfrak{p}) - 1$ over k. Further, given a noether filtration f on Q, the set of valuations associated with f contains one, v say, such that its extension $v^\wedge$ takes the value ∞ on a given minimal prime ideal $\mathbf{p}$ of Q^. In chapter 7 it is sufficient to restrict A to be a domain, and for v to be an ideal valuation on A, it is necessary and sufficient for there to exist a finitely generated extension B of A contained in the field of fractions of A with the following two properties: first, that $v(x) \geq 0$ on B and secondly, that the centre $\mathbf{p}$ of v on B has height 1. Here, by the centre of v on B, we mean the prime ideal $\mathbf{p}$ on which v takes positive values. Using these two characterisations of ideal valuations, we can describe the ideal valuations of a finitely generated extension B of A in terms of those of A. Further, it is possible to use ideal valuations to give new proofs of a number of results of the type of the altitude inequality or concerning chains of prime ideals. This is done in chapter 7.

The next two chapters are concerned with multiplicities. In chapter 8, if f is a noether filtration on a local ring $(Q,\mathfrak{m},k,d)$ with radical $\mathfrak{m}$, we associate with f an additive function e(f,M) on the category FG(Q) of finitely generated Q-modules. This is very closely related to the ordinary multiplicity function associated with an ideal. The theory is developed in the short chapter 8 by means of Koszul Complexes. More important from the point of view of these notes is the degree function d(f,M,x) introduced in chapter 9. This depends not only on f, but on an element x of Q satisfying the condition that $\dim(Q/xQ) = d - 1$, and, for fixed f, x is again an additive function of M. It is defined by

$$d(f,M,x) = e(f_x,M/xM) - e(f_x, (0:x)_M)$$

where f_x is the filtration on Q/xQ defined by taking $f_x(y) = \mathrm{Max}\,(f(y'))$, where y' ranges over the inverse images of y under the map $Q \to Q/xQ$, and M/xM, $(0:x)_M$ are considered as (Q/xQ)-modules. The main object of these two chapters is to prove a degree formula

$$d(f,M,x) = \Sigma\delta(v)L_v(M)d(f,v)v(x),$$

where v ranges over the valuations associated with f such that the residue field of v has transcendence degree d−1, $\delta(v)$ is the length of the primary component of zero in Q^ corresponding to the prime ideal $\mathbf{p}$ of Q^ on which v^ takes the value infinity, and $L_v(M)$ is the length of the module $M_{\mathbf{p}(v)}$ over the artinian ring $Q_{\mathbf{p}(v)}$, where $\mathbf{p}(v)$ is the minimal prime ideal of Q on which v takes the value ∞. The numbers d(f,v) are positive rational numbers. Chapter 9 concludes with a proof, using the degree formula, of the theorem that, if Q is quasi-unmixed (that is, $\dim(Q\hat{}/\mathbf{p}) = d$ for all minimal primes $\mathbf{p}$ of $Q^{\wedge}$), and f,g are two noether filtrations with radical $\mathbf{m}$ such that $g(x) \geq f(x)$ for all x, then f,g are equivalent if and only if $e(f,Q) = e(g,Q)$.

Now we turn to the last three chapters. We have already remarked that these chapters are separate from the first 9, and that the key idea underlying them is that of general elements of ideals. If $(Q,\mathbf{m},k,d)$ is a local ring and J is an ideal of Q with a basis $a_1,\ldots,a_m$, then a natural way of defining a general element x of J is to take $x = \Sigma X_i a_i$, where $X_1,\ldots,X_m$ are indeterminates over Q. Further, if we localise $Q[X_1,\ldots,X_m]$ at $\mathbf{m}[X_1,\ldots,X_m]$, we keep within the theory of local rings. However, in situations where we have to consider several general elements of either the same or different ideals, this means that we have to adjoin more indeterminates. Eakin and Sathaye[1976] overcame this difficulty by adjoining a countable set of indeterminates $X_1,X_2,\ldots$ and localising at $\mathbf{m}[X_1,X_2,\ldots]$, using the resulting ring with considerable success. It is this idea we follow in these three chapters, the resulting ring being termed the *general extension* of Q and denoted by Q_g. Q_g can also be considered as the union of the rings Q_N obtained by localising $Q[X_1,\ldots,X_N]$ at $\mathbf{m}[X_1,\ldots,X_N]$. In these notes considerable use is made of the fact that Q_g is noetherian. This is a particular case of a general result of Grothendieck which appears as Proposition 1 in the appendix to chapter 9 of [BAC], but an *ad hoc* proof of the fact that Q_g is noetherian is given in chapter 10. Another aspect of Q_g is that it is a regular extension of Q, which implies that any reasonable condition imposed on Q is almost certainly also satisfied by Q_g. This is verified in a number of important cases in chapter 10. However, the main object of chapter 10 is twofold. First we study the relationship between a prime ideal $\mathbf{P}$ of Q_g and the prime ideal $\mathbf{p} = \mathbf{P} \cap Q$ of Q. The basic result is Theorem 10.23 of chapter 10. Secondly we consider $\mathbf{m}$-valuations. An $\mathbf{m}$-valuation v on Q is a valuation non-negative on Q, positive on $\mathbf{m}$, and taking the value ∞ on some prime ideal $\mathbf{p}$ of Q, not necessarily minimal. We require the further restriction on v that its residue field K_v be finitely generated over k. Finally we say that v is *good* if

$$\text{trans.deg}_k K_v = \dim(Q/\mathbf{p}) - 1.$$

The results we require on $\mathfrak{m}$-valuations on Q, Q_g are set out in the third part of chapter 10.

In chapter 11 we now come to general elements. We will only consider here the simplest case, that of a general element of an ideal J. The definition already given is taken here as the definition of a standard general element x of J. A general element x of J is then any element of the form T(x), where T is an automorphism of Q_g over Q. This definition turns out to be independent of the choice of the particular set of generators $a_1,\ldots,a_m$ of J we start with. It is not too difficult to extend the same idea to consider independent sets of general elements $x_1,\ldots,x_s$ of a set of ideals $J_1,\ldots,J_s$ and to remark upon results such as the ideal $Q\cap(x_1Q_g+\ldots+x_sQ_g)$ depending only on $J_1,\ldots,J_s$.

Now we turn to the final chapter where the object is to generalise the degree formula of chapter 9, although filtrations play no part in the generalisation. We start with a definition. A set of ideals $J_1,\ldots,J_r$ of Q is said to be *independent* if a set of independent general elements $x_1,\ldots,x_r$ of these ideals is a sub-set of a set of parameters of Q_g. If r = d, then we can define a mixed multiplicity $e(Q|\,\mathbf{J}\,|M)$ as the multiplicity $e(Q_g|\,x_1,\ldots,x_d\,|M\otimes_Q Q_g)$, as introduced by D.J. Wright and described in detail in Northcott[LRMM]. This depends only on the set of ideals $\mathbf{J}=(J_1,\ldots,J_d)$, and not on the set of independent general elements $x_1,\ldots,x_d$ chosen. If the ideals of $\mathbf{J}$ are all $\mathfrak{m}$-primary, this is Teissier's mixed multiplicity, but it is defined for some other sets of ideals $\mathbf{J}$. The function $e(Q|\,\mathbf{J}\,|M)$ is symmetric in the ideals $J_1,\ldots,J_d$, takes non-negative integer values and has the nice property that, if we write $\mathbf{J}'$ for the set $(J_1,\ldots,J_{d-1})$,

$$e(Q|\,\mathbf{J}',J_dK_d\,|M)=e(Q|\,\mathbf{J}',J_d\,|M)+e(Q|\,\mathbf{J}',K_d\,|M).$$

The general degree formula now arises as follows. We consider a set of d-1 independent ideals $\mathbf{J}'$ as above. If J is any $\mathfrak{m}$-primary ideal, the set $(\mathbf{J}',J)$ is also independent, and hence for any M, $e(Q|\,\mathbf{J}',J\,|M)$ can be considered as a function on the set of $\mathfrak{m}$-primary ideals of Q with non-negative integer values. In its simplest form, the degree formula is a formula

$$e(Q|\,J',J\,|M) = \Sigma a(J';M;v)v(J)$$

the sum being over all good $\boldsymbol{m}$-valuations on Q. The existence of such a formula implies the uniqueness of the coefficients a(J';M;v) as a result of a theorem proved in chapter 10. We refer the reader to chapter 12 for more specific information concerning the coefficients a(J';M;v). We simply note here that the values of a(J';M;v) are non-negative integers, and, as a function of M, they are additive functions on FG(Q).

1. GRADED RINGS AND MODULES

1. Definitions and Samuel's Theorem.

Throughout these notes the rings considered will, unless otherwise stated, be commutative, and will possess an identity element 1. If A is a ring, an A-module M will always be assumed to be unitary, i.e., 1.x = x for all x in M. Note that scalars are written on the left.

DEFINITION. *A ring* G *will be said to be graded if the additive group of* G *is the direct sum of additive groups* G_r, *where* r *runs over all integers, and*

$$G_{r+s} \supseteq G_r.G_s$$

for all r,s.

Let G *be a graded ring and let* M *be a* G*-module. Then* M *will be termed a graded* G*-module if* M *is the direct sum of additive groups* M_r *such that*

$$M_{r+s} \supseteq G_r.M_s$$

for all r,s.

If N *is a sub-module of a graded module* M, *then* N *is termed a graded sub-module of* M *if* N *is the direct sum of the additive groups* $N \cap M_r$. *An ideal* I *of a graded ring* G *is a graded ideal if it is a graded sub-module of* G.

In what follows we will consider a graded ring G as given, and all G-modules, sub-modules and ideals will be considered to be graded unless otherwise stated. Further, we will refer to an element of the additive group G_r (or M_r) as *homogeneous* (of degree r). We will also denote the sub-ring of G consisting of all sums of homogeneous elements of degree ≥ 0 by G^+ and the sub-ring consisting of all sums of homogeneous elements of degree ≤ 0 by G^-.

DEFINITION. *A graded ring* G *will be said to be a graded noetherian ring if all (graded) ideals of* G *are finitely generated.*

The theorem following will be referred to as Samuel's Theorem, since the author first met it in Samuel[1953].

THEOREM 1.11 *If G is a graded ring, then the following statements about G are equivalent:*

i) G *is a graded noetherian ring,*

ii) G_0, G^+ *and* G^- *are graded noetherian rings, and for all* r, G_r *is a finite* G_0*-module,*

iii) G_0 *is a noetherian ring and there exist homogeneous elements* $x_1,\dots,x_r$ *of positive degree and homogeneous elements* $y_1,\dots,y_s$ *of negative degree such that* $G^+ = G_0[x_1,\dots,x_r]$ *and* $G^- = G_0[y_1,\dots,y_s]$.

i) implies ii). Let I be an ideal of G_0. Then IG can be generated by a finite set of elements which can be chosen from I. Clearly these elements generate I as an ideal of G_0. Hence I is finitely generated. Since I is an arbitrary ideal of G_0, G_0 is noetherian. Next, consider the ideal of G generated by the elements of G_r. This ideal is generated by a finite set of elements of G_r, which clearly generate G_r as a G_0-module. Now let I be a graded ideal of G^+. Again IG is generated by a finite set of homogeneous elements of I. Let m be the maximum of the degrees of these elements. Then the ideal of G^+ generated by these elements contains all elements of I of degree $\geq$m. Further, $I \cap G_r$ is a finitely generated G_0-module for $r < m$. Hence, if we add to our set of generators sets of generators of $I \cap G_r$ for $r = 0,\dots,m-1$, we obtain a finite set of generators of I. Hence G^+ is noetherian. The proof that G^- is noetherian is precisely similar.

ii) implies iii). Let g^+ denote the ideal of G^+ generated by all homogeneous elements of G of positive degree. This ideal has a finite set of homogeneous generators $x_1,\dots,x_r$. Let x_i have degree $d_i > 0$. Let z be an element of G^+ of degree n. We prove by induction on n that z is expressible as a polynomial in $x_1,\dots,x_r$ with coefficients in G_0. This is obvious if $n = 0$. If $n > 0$ then we can express z in the form

$$z = \Sigma\, x_i z_i$$

where z_i is homogeneous of degree $n - d_i < n$, and hence is expressible as a polynomial in $x_1,\dots x_r$. Hence z is expressible as a polynomial over G_0 in $x_1,\dots,x_r$. This completes the proof of the statement as regards G^+ and the proof of the statement regarding G^- is precisely similar.

iii) implies i). The hypotheses of iii) imply that $G = G_0[x_1,\dots,x_r;y_1,\dots,y_s]$ and hence, by the Hilbert Basis Theorem, all ideals of G (even non-graded ideals) are finitely generated. Hence G is a graded noetherian ring.

For convenience of reference later, we extract from the above proof the following as a corollary.

COROLLARY. *Let* G *be a noetherian graded ring, let* g^+ *denote the ideal of* G^+ *generated by homogeneous elements of degree* >0, *and let* g^- *denote the ideal of* G^- *generated by homogeneous elements of degree* <0. *Then if* $x_1,\dots,x_r$ *is a set of homogeneous generators of* g^+, *and* $y_1,\dots,y_s$ *a set of homogeneous generators of* g^-, *these elements can be taken as the elements* $x_1,\dots,x_s : y_1,\dots,y_s$ *of* iii).

This is proved above, under ii) implies iii).

For simplicity, we now restrict ourselves to graded rings such that all homogeneous elements other than zero have degree ≥0. If G satisfies this condition, we shall say that G is *positively graded.* If G is positively graded, an ideal J of G will be termed *irrelevant* if it contains all homogeneous elements of sufficiently high degree.

THEOREM 1.12. *Let* G *be a positively graded noetherian ring, and let* $x_1,\dots,x_r$ *be homogeneous elements of* G *of positive degree. Then the following conditions on* $x_1,\dots,x_r$ *are equivalent,*

i) *The ideal* J *generated by* $x_1,\dots,x_r$ *is irrelevant.*

ii) *If* H *is the graded sub-ring* $G_0[x_1,\dots,x_r]$ *of* G, *then* G *is a finite* H*-module.*

i) implies ii). Suppose that all homogeneous elements of degree ≥n belong to J. Then by induction on N−n, we can prove, as in the proof of ii) implies iii) in the last theorem, that every homogeneous element z of G of degree N ≥ n can be expressed in the form

$$z = \Sigma f_i u_i$$

where f_i is a polynomial over G_0 in $x_1,\dots,x_r$ and u_i is a homogeneous element of G of degree <n. Since $G_1,\dots,G_{n-1}$ are all finitely generated G_0-modules, the result follows.

ii) implies i). Let $u_1,\dots,u_k$ be a set of homogeneous elements of G such that

$$G = Hu_1 + \dots + Hu_k$$

and let n be the maximum of the degrees of $u_1,...,u_k$. Then any element of G of degree $>n$ is contained in the ideal $(x_1,...,x_r)$.

COROLLARY. *If* G *is a positively graded noetherian ring, an ideal* J *of* G *is irrelevant if and only if* J *contains a power of* g^+, *the ideal of* G *generated by all elements of positive degree.*

If J is irrelevant, then J contains a power of g^+ whether G is noetherian or not. Conversely, suppose that $x_1,...,x_n$ is a set of generators of g^+ and that the maximum of the degrees of $x_1,...,x_n$ is m. Then if r is a positive integer and $N > rm$, any element z of G_N is a linear combination over G_0 of monomials in $x_1,...,x_n$ whose degrees are $\geq r$. It follows that z belongs to $(g^+)^r$. Hence $(g^+)^r$ is irrelevant for any r and so is any graded ideal containing a power of g^+.

For our final two theorems in this section, we need neither the restriction to noetherian graded rings nor the restriction to positively graded rings.

THEOREM 1.13. *Let* G *be a graded ring,* M *be a graded* G*-module, and* I *be a finitely generated ideal of* G *(not assumed to be graded). Suppose that* M *contains a non-zero element* m' *such that* Im' = 0. *Then* M *contains a non-zero homogeneous element* m *such that* Im = 0. *Consequently, if* I' *is the smallest graded ideal of* G *containing* I, *then* I'm = 0.

Let f(1),...,f(r) be a set of generators of I. Then each element f(i) can be expressed uniquely in the form Σf(i,j), where the elements f(i,j) are non-zero homogeneous elements of G whose degrees increase strictly as j increases. It follows that I' is the ideal generated by the elements f(i,j). Further, we can also write

$$m' = m(1) + ... + m(s)$$

where the elements m(k) are non-zero homogeneous elements of M whose degrees increase strictly with k. We may also suppose that m' is chosen so that no non-zero element m" of M which is the sum of less than s homogeneous elements satisfies Im" = 0.

Now suppose that, with this restriction on m', we have proved for a particular i that f(i,j)m' = 0 for j = 1,...,h. Then it follows that f(i,h+1)m(1) = 0. Hence m" = f(i,h+1)m' is the sum of fewer than s homogeneous elements and satisfies Im" = 0. Hence m" = 0, and, proceeding in this way, we can prove that f(i,j)m' = 0 for all i,j. We now take $m = m(1)$.

The next theorem is the prime avoidance theorem.

THEOREM 1.14. *Let* G *be a graded ring,* H *a graded sub-ring of* G *and* $\mathbf{p}_1,\ldots,\mathbf{p}_s$ *prime ideals of* G *with the property that, for each* i, *there is a homogeneous element* x_i *of* H *of positive degree such that* x_i *does not belong to* $\mathbf{p}_i$. *Then there exists a homogeneous element* x *of positive degree in* H *which belongs to none of the ideals* $\mathbf{p}_i$.

We use induction on s, the case $s = 1$ being immediate. We can assume for each i that there exists a homogeneous element y_i of H of positive degree which is not in any of the prime ideals $\mathbf{p}_j$ for $j \neq i$. If for any i, y_i is not in $\mathbf{p}_i$, we are done. Otherwise, first replace each of the elements x_i, y_i by suitable powers so that the elements x_i, y_i all have the same degree. Then

$$x = \Sigma_i y_1 \ldots y_{i-1} x_i y_{i+1} \ldots y_s$$

is the required element.

COROLLARY. Let G *be a graded ring,* I *be an ideal of* G *and* $\mathbf{p}_1,\ldots,\mathbf{p}_s$ *be prime ideals of* G *with the property that, for each* i, *there is a homogeneous element* x_i *of positive degree in* I *but not in* $\mathbf{p}_i$. *Then there is a homogeneous element* x *of positive degree in* I *but not in any of the prime ideals* $\mathbf{p}_i$.

Take H to be $G_0 + (G_1 \cap I) + \ldots + (G_r \cap I) + \ldots$ in the theorem.

2. Rappel on Koszul complexes.

We recall here the results on Koszul complexes we will require in this section, particularly in the context of graded rings. We suppose that $\mathbf{x} = x_1,\ldots,x_q$ is a set of homogeneous elements of a graded ring G, with degrees $f_1,\ldots,f_q$. For each sub-set S of the set $1,\ldots,q$, we now define a symbol u(S). If p is an integer satisfying $0 \leq p \leq q$, we define $K_p(\mathbf{x},G)$ to be the free G-module with generators u(S) for those sets S which contain p elements. We grade $K_p(\mathbf{x},G)$ by defining the degree of xu(S), where x is a homogeneous element of G of degree d, to be $d + \Sigma f_i$, where the sum is over those i which belong to S. Next we define the differential d. If $S = \{s_1,\ldots,s_p\}$, where $s_1 < s_2 < \ldots < s_p$, we denote by S_k the set obtained by omitting s_k from S and also let $x(i) = x_i$. Then we define the map

$$d:K_p(\mathbf{x},G) \longrightarrow K_{p-1}(\mathbf{x},G)$$

by

$$d(u(S)) = x(s_1)u(S_1) - x(s_2)u(S_2) + \ldots + (-1)^{p-1}x(s_p)u(S_p)$$

and linearity over G. Then $d^2 = 0$ and d preserves degrees. We have thus defined the Koszul Complex:

$$K(\mathbf{x},G)\colon 0 \to K_q(\mathbf{x},G) \xrightarrow{d} K_{q-1}(\mathbf{x},G) \xrightarrow{d} \ldots \xrightarrow{d} K_1(\mathbf{x},G) \xrightarrow{d} K_0(\mathbf{x},G) \to 0.$$

If M is a graded G-module, we define $K(\mathbf{x},M) = K(\mathbf{x},G)\otimes_G M$. Let $H_i(\mathbf{x},M)$, $i=0,\ldots,q$, denote the homology modules of $K(\mathbf{x},M)$. These are then graded G-modules, annihilated by the ideal $(x_1,\ldots,x_q)$. In particular, if G is noetherian and M is finitely generated, they are also finitely generated. Finally, we have

$$H_0(\mathbf{x},M) = M/(x_1,\ldots,x_q)M \;;\; H_q(\mathbf{x},M) = (0:(x_1,\ldots,x_q))_M(\Sigma f_i)$$

where the convention is adopted that, if M is a graded G-module, M(n) denotes the module derived from M by increasing the degrees of all homogeneous elements by n.

To conclude this elementary review of Koszul Complexes, we recall that if

$$0 \longrightarrow L \longrightarrow M \longrightarrow N \longrightarrow 0$$

is a short exact sequence of G-modules, then we have a long exact sequence

$$0 \longrightarrow H_q(\mathbf{x},L) \longrightarrow H_q(\mathbf{x},M) \longrightarrow H_q(\mathbf{x},N) \longrightarrow H_{q-1}(\mathbf{x},L) \longrightarrow \ldots$$

$$\ldots \longrightarrow H_{i+1}(\mathbf{x},N) \longrightarrow H_i(\mathbf{x},L) \longrightarrow H_i(\mathbf{x},M) \longrightarrow H_i(\mathbf{x},N) \longrightarrow H_{i-1}(\mathbf{x},L) \longrightarrow \ldots$$

$$\ldots \longrightarrow H_1(\mathbf{x},N) \longrightarrow H_0(\mathbf{x},L) \longrightarrow H_0(\mathbf{x},M) \longrightarrow H_0(\mathbf{x},N) \longrightarrow 0\,.$$

3. Additive functions on modules.

To commence this section, we suppose that A is a noetherian ring and that FG(A) denotes the category of finitely generated A-modules. We are concerned with functions f(M) defined on FG(A), with values in some additive abelian group H, which satisfy the condition that, if

$$0 \longrightarrow L \longrightarrow M \longrightarrow N \longrightarrow 0$$

is a short exact sequence in FG(A), then

$$f(L) - f(M) + f(N) = 0.$$

We refer to such a function as an *additive* function. We now take as known the existence of an additive group Gr(A), the Grothendieck group of the category FG(A), and an additive function μ as above with values in Gr(A), such that, if f is an additive function with values in H, then there is a unique homomorphism g: Gr(A) $\rightarrow$ H satisfying $g\mu = f$. In order to discuss the existence of such functions, we first need the following lemma.

LEMMA 1.31. *Let* M *be a finitely generated* A-*module. Then there exists a sequence of sub-modules*

$$M = M_0 \supset M_1 \supset \ldots \supset M_n = (0)$$

such that M_{i-1}/M_i *is isomorphic to* $A/\mathbf{p}_i$ *for some prime ideal* $\mathbf{p}_i$ *of* A *and* $i = 1,\ldots,n$.

We use noetherian induction. Since the result is trivial for the zero module, we can assume the existence of a maximal sub-module N of M for which the lemma holds. It then follows that, if N ≠ M, the quotient module M/N has no sub-module isomorphic to $A/\mathbf{p}$ for any prime ideal $\mathbf{p}$ of A. We now show that this leads to a contradiction. Choose u in M/N such that J = Ann(Au) is maximal in the set of annihilator ideals of non-zero elements of M/N. Then Au is isomorphic to A/J. Then J must be prime. For suppose that a, b are elements of A such that $ab \in J$ but a does not belong to J. Then au ≠ 0, and its annihilator contains J, and hence equals J by the maximality of J. But the annihilator of au contains b and hence b belongs to J. Hence J is prime, and therefore N = M, proving the result.

We will refer to the sequence of sub-modules $M_0,\ldots,M_n$ as a *prime filtration* of M. One consequence of the existence of prime filtrations is that an additive function f is determined by its values on the modules $A/\mathbf{p}$ where $\mathbf{p}$ ranges over all prime ideals of A. For the last lemma shows that $f(M) = \Sigma f(A/\mathbf{p}_i)$, the sum being for i = 1 to n.

In general, the number of occurrences of $A/\mathbf{p}$ as a factor M_{i-1}/M_i in a prime filtration will depend on the filtration. However, if we put a restriction on $\mathbf{p}$, this is not the case as the following lemma shows.

LEMMA 1.32. i) *If* A *is an artinian ring with prime ideals* $\mathbf{p}_1,\ldots,\mathbf{p}_r$, *then the number of occurrences of* $A/\mathbf{p}_i$ *as a factor in a prime filtration of a finitely generated* A-*module* M *is independent of the filtration chosen.*

ii) *If* A *is a noetherian ring, and* $\mathbf{p}$ *is a minimal prime ideal of* A, *the number of occurrences of* $A/\mathbf{p}$ *as a factor of a prime filtration of a finitely generated* A-*module* M *is independent of the filtration.*

i) Since A is an artinian ring, the modules $A/\mathbf{p}_i$ are the simple A-modules, and the statement is therefore a consequence of the Jordan-Holder Theorem.

ii) Let $\mathbf{p}$ be a minimal prime ideal of A. The functor $M \longrightarrow M_{\mathbf{p}}$ is an exact functor from the category of finitely generated A-modules to the category of finitely generated $A_{\mathbf{p}}$-modules. Further, $(A/\mathbf{p}')_{\mathbf{p}} = 0$ if $\mathbf{p}' \neq \mathbf{p}$. Hence, localisation at $\mathbf{p}$ as applied to a prime filtration of M deletes all factors not isomorphic to $A/\mathbf{p}$ and replaces those factors isomorphic to $A/\mathbf{p}$ by $A_{\mathbf{p}}/\mathbf{p}A_{\mathbf{p}}$ yielding a filtration of the $A_{\mathbf{p}}$-module $M_{\mathbf{p}}$. But $A_{\mathbf{p}}$ is an artinian ring with just one prime ideal $\mathbf{p}A_{\mathbf{p}}$. Hence all prime filtrations of $M_{\mathbf{p}}$ have the same length, and this equals the number of occurrences of $A/\mathbf{p}$ in any prime filtration of M.

The above lemma shows that, if $\mathbf{p}$ is a minimal prime ideal of A, the number of occurrences of $A/\mathbf{p}$ as a factor in a prime filtration of a finitely generated A-module M is an additive function $f_{\mathbf{p}}$ on FG(A) with values in the ring of integers Z. Further, $f_{\mathbf{p}}(M) \geq 0$ for all A-modules M. This latter condition on additive functions to the integers (or, more generally, to ordered additive groups) almost reduces us to the functions $f_{\mathbf{p}}$, as the following lemma shows.

LEMMA 1.33. *If* f *is an additive function from* FG(A) *into* Z *such that* $f(M) \geq 0$ *for all* M, *then*

$$f(M) = \Sigma\, f(A/\mathbf{p})f_{\mathbf{p}}(M)$$

for all M, *the sum being over the minimal prime ideals* $\mathbf{p}$ *of* A. *Further,* $f(A/\mathbf{p}) \geq 0$ *for all such* $\mathbf{p}$.

We have to show that $f(A/\mathbf{p}') = 0$ if $\mathbf{p}'$ is not minimal. If $\mathbf{p}'$ is not minimal, then there is a minimal prime ideal $\mathbf{p}$ of A and an element x of A not in $\mathbf{p}$ such that $\mathbf{p}+xA$ is contained in $\mathbf{p}'$. Then the exact sequence

$$0 \longrightarrow A/\mathbf{p} \xrightarrow{\ \mu_x\ } A/\mathbf{p} \longrightarrow A/(\mathbf{p}+xA) \longrightarrow 0$$

where μ_x denotes multiplication by x, shows that $f(A/\mathbf{p}+xA) = 0$ for any additive function f. But we also have an exact sequence

$$0 \longrightarrow \mathbf{p}'/(\mathbf{p}+xA) \longrightarrow A/(\mathbf{p}+xA) \longrightarrow A/\mathbf{p}' \longrightarrow 0$$

which gives

$$f(\mathbf{p}'/(\mathbf{p}+xA)) + f(A/\mathbf{p}') = f(A/(\mathbf{p}+xA)) = 0$$

and since both terms on the left-hand side are, by hypothesis, non-negative, it follows that $f(A/\mathbf{p}') = 0$. Hence f must be a linear combination of the additive functions $f_{\mathbf{p}}$, and the result now follows.

We now replace A by a graded noetherian ring G, and consider the category of finitely generated graded G-modules which we will denote by FG(G). In considering additive functions, we note that if

$$0 \longrightarrow L \text{---} f \rightarrow M \text{---} g \rightarrow N \longrightarrow 0$$

is an exact sequence, then it is implicitly assumed that f,g preserve degrees. It follows that if $n \neq 0$, M and M(n) have different images in Gr(G). It also means that Gr(G) has additional structure, in that it is a $Z[X,X^{-1}]$-module, if we define $X.\mu(M) = \mu(M(1))$. The results above remain true on the whole in the graded case if we make some modifications. First, Lemma 1.31 remains true if we only require that M_{i-1}/M_i be isomorphic to $(A/\mathbf{p})(n)$ for some n, and make the corresponding change in the definition of a prime filtration. Lemma 1.32 remains true if we replace the number of occurrences of $A/\mathbf{p}$ by the total number of occurrences of $(A/\mathbf{p})(n)$ for all n. However Lemma 1.33 is no longer valid, since the element x need not be of degree 0, and hence the morphism μ_x does not preserve degrees.

In the next section we shall be concerned with additive functions on the category FG(G) of graded and finitely generated modules over a positively graded noetherian ring G which arise as follows. Suppose that f is an additive function on $FG(G_0)$ into an abelian group H. Then each module M in FG(G) determines an element $h_n(f,M)$ of H for each integer n by $h_n(f,M) = f(M_n)$, and the functions $h_n(f, .)$ are clearly additive functions from FG(G) to H for each n. Note that for each M, $h_n(f,M) = 0$ if n is large and negative, since we are assuming that G is positively graded, implying that $M_n = 0$ if n is large and negative. We will consider all these functions together by introducing the formal power series $h(f,M,w) = \Sigma h_n(f,M)w^n$. Note that this is a formal power series in w with coefficients in H, and further that the coefficients of large negative powers in h(f,M,w) are zero. Then h(f,M,w) is an additive function from FG(G) to an additive group H<w> consisting of all formal power series in w with coefficients in H, negative powers of w being allowed, but for each element of H<w> there is an integer m depending on the element such that the coefficient of w^n is zero if $n < m$. H<w> is a module over the ring of formal power series Z<w> defined in a similar manner.

4. The Hilbert series of a graded module.

In this section G will be a positively graded noetherian ring. We now use the end of the last section to make the following definition.

DEFINITION. *If* M *is a finitely generated* G-*module, the formal power series* h(f,M,w) *will be termed the Hilbert series of* M *with respect to* f. *If* H *is* $Gr(G_0)$ *and* f *is the associated additive function* μ, *then we will write* h(M,w) *in place of* $h(\mu,M,w)$ *and refer to this as the Hilbert series of M.*

Our object in this section is to give a formula for h(M,w). We commence by making a number of simplifications. Suppose that $y_1,\ldots,y_d$ is a set of homogeneous elements of G of degrees $f_1,\ldots,f_d > 0$ which generate an irrelevant ideal of G, or, what is the same thing, satisfy the condition that G is a finitely generated $G_0[y_1,\ldots,y_d]$-module. Now let f denote the l.c.m. of $f_1,\ldots,f_d$ and let x_i denote y_i raised to the power f/f_i. Then G, and hence M, is still a finite $G_0[x_1,\ldots,x_d]$-module. Hence in calculating h(M,w), we can replace G by $G_0[x_1,\ldots,x_d]$, i.e., we can assume that G is generated over G_0 by elements of the same positive degree f. We therefore make this assumption. We now make a further simplification. Let i be an integer satisfying $0 \le i < f$ and consider the sub-module $M^{(i)}$ of M consisting of sums of homogeneous elements of M whose degrees equal i modulo f. (Note that this is a sub-module and is finitely generated.) We now take two further steps. First we divide the degrees of homogeneous elements of G by f. (Note that the earlier simplifications have ensured that the degrees of non-zero elements are multiples of f.) Secondly, we decrease degrees in $M^{(i)}$ by i and divide degrees by f. We denote the new ring by G/f and the new module by $M^{(i)}/f$. Then the following is almost immediate.

$$h(M,w) = \Sigma\, w^i h(M^{(i)}/f, w^f) \qquad (1.40)$$

the sum being over i = 0 to f-1. Hence the calculation of h(M,w) is reduced to the case where G is generated over G_0 by elements of degree 1, and we will now restrict ourselves to this case.

Our main tool will be the Koszul Complex, and we make use of the following general result.

LEMMA 1.41. *Let*

$$C: 0 \longrightarrow C_n \xrightarrow{d_n} C_{n-1} \xrightarrow{d_{n-1}} \cdots C_2 \xrightarrow{d_2} C_1 \xrightarrow{d_1} C_0 \longrightarrow 0$$

be a complex of finitely generated A-modules, and let f be an additive function on FG(A). *Then*

$$\Sigma(-1)^i f(C_i) = \Sigma(-1)^i f(H_i(C))$$

both sums being from $i = 0$ *to* n.

Let $Z_i = \text{Ker } d_i$ and $B_i = \text{im } d_{i+1}$. Then we have exact sequences

$$0 \longrightarrow B_i \longrightarrow Z_i \longrightarrow H_i(C) \longrightarrow 0 \quad \text{and} \quad 0 \longrightarrow Z_i \longrightarrow C_i \longrightarrow B_{i-1} \longrightarrow 0.$$

If we now apply f to these sequences we obtain

$$f(H_i(\mathbf{C})) = f(Z_i) - f(B_i) \quad \text{and} \quad f(C_i) = f(Z_i) + f(B_{i-1}).$$

Multiplying each by $(-1)^i$ and summing each from $i = 0$ to n then yields the required result, since $B_n = (0)$.

Now we turn to the Koszul complex $K(\mathbf{x},M)$ where $\mathbf{x} = x_1,\dots,x_d$ and $G = G_0[x_1,\dots,x_d]$. We assume that $x_1,\dots,x_d$ all have degree 1.

Now the modules $H_i(x,M)$ are annihilated by $(x_1,\dots,x_d)$ and are finitely generated. follows that, for each i, $H_i(\mathbf{x},M)_n = 0$ for n large and hence $h(H_i(\mathbf{x},M),w)$ is a finite sum, i.e., is a polynomial in w,w^{-1}. Hence the same is true of $\Sigma(-1)^i h(H_i(\mathbf{x},M),w)$, which we will denote by $\chi(M,w)$. We now apply Lemma 1.41 with $A = G$ and $f = \mu$ to note that

$$\chi(M,w) = \Sigma(-1)^i h(K_i(\mathbf{x},M),w)$$

the sum being for $i = 0$ to d. But $K_i(\mathbf{x},M)$ is the sum of copies of $M(i)$, one for each sub-set containing i elements of 1,...,d, and hence $\binom{d}{i}$ in number, where $\binom{d}{i}$ denotes the binomial coefficient $d!/(d-i)!i!$. But

$$h(M(i),w) = w^i h(M,w)$$

and hence the right-hand side of the above equation is

$$(1-w)^d\, h(M,w).$$

Equating the two sides, we obtain

$$h(M,w) = (1-w)^{-d}x(M,w)$$

where $x(M,w)$ is a finite sum, and therefore of the form $w^{-k}p(M,w)$, for some integer k with $p(M,w)$ a polynomial in w. Note that $(1-w)^{-d}$ has to be interpreted as a formal power series in w in the obvious way. Combining this with (1.40) we obtain our general result, which we state as the following theorem.

THEOREM 1.42. *Let G be a graded noetherian ring which is positively graded and let* M *be a finitely generated graded* G-*module. Suppose there exist homogeneous elements* $x_1,\dots,x_d$ *in G of positive degrees* $f_1,\dots,f_d$ *generating an irrelevant ideal. Then if* f *is the l.c.m. of* $f_1,\dots,f_d$, *we can write*

$$h(M,w) = w^{-k}(1-w^f)^{-d}q(M,w)$$

where $q(M,w)$ *is a polynomial with coefficients in* $Gr(G_0)$, *and* k *is an integer.*

COROLLARY. *Let* G, M, $x_1,\dots,x_d$, $f_1,\dots,f_d$ *be as above, and let* $a(.)$ *be an additive function on* $FG(G_0)$ *with values in an additive abelian group* H. *Then*

$$h(a,M,w) = w^{-k}(1-w^f)^{-d}q(a,M,w),$$

where $q(a,M,w)$ *is a polynomial in* w *with coefficients in* H.

We can find a homomorphism $g:Gr(G_0) \to H$ such that $a(N) = g(\mu(N))$. Applying g to the equation above we obtain the required result with $q(a,M,w) = g(q(M,w))$.

For our next development we require a modification of the expression given for $h(a,M,w)$. First we express $q(a,M,w)$ in the form $(1-w^f)^m . r(a,M,w)$ where m is the largest integer such that $(1-w^f)^m$ divides $q(a,M,w)$ without remainder. Hence we now have

$$h(a,M,w) = w^{-k}(1-w^f)^{m-d}.r(a,M,w).$$

Note that if $m \geq d$, this implies that $a(M_n) = 0$ for n large. We now express $r(a,M,w)$ in the form

$$r(a,M,w) = r_0(a,M,w) + (1-w^f)r_1(a,M,w) + \dots + (1-w^f)^i r_i(a,M,w) + \dots$$

where $r_i(a,M,w)$ is a polynomial of degree $<f$. We now consider the value of $a(M_n)$ for n large. We have already dealt with the case $m \geq d$. Hence we will assume that $m < d$ and we will write $n + k = rf + j$ where $0 \leq j < f$. Let a_{ij} denote the coefficient of w^j in $r_i(a,M,w)$. Then, if n is large, the coefficient of w^n in $h(a,M,w)$ is

$$a_{0j}\binom{r+d-m-1}{d-m-1} + \dots + a_{ij}\binom{r+d-m-i-1}{d-m-i-1} + \dots \quad .$$

For each j, this is a polynomial $H_j(a,M,r)$ in r for r large.

We now turn to dimension theory. However, since for graded rings the definitions of dimension we shall use differ from the geometrical notion of dimension, we will use the term *spread* instead of dimension. For our first definition, we suppose that a is an additive function on the category of finitely generated G_0-modules to a partially ordered group satisfying the condition that $a(M) \geq 0$ for all modules M. We now define the *a-spread* s(a,M) of a module M to be the maximum of the degrees of the polynomials $H_j(a,M)$ increased by 1, with the proviso that, if all the polynomials $H_j(a,M)$ are zero, then s(a,M) = 0. Note that if s(a,M) = 0, then $a(M_n) = 0$ for all large n. We will, in fact, make little use of s(a,M) in these notes, and then only when G_0 is a field k. In this case the only possible candidate for a is essentially $a = \dim_k$ which takes integer values and a more convenient definition of s(a,M) is that s(a,M) is the least integer s such that

$$n^{-s}a(M_n) \to 0 \quad \text{as} \quad n \to \infty.$$

Note however that, in general, s(a,M) has the properties that

a) s(a,M(r)) = s(a,M) for all integers r,

b) if

$$0 \longrightarrow M' \longrightarrow M \longrightarrow M'' \longrightarrow 0$$

is an exact sequence of graded G-modules, then

$$s(a,M) = \mathrm{Max}(s(a,M'),s(a,M'')).$$

We now consider a second approach to the notion of spread. This is contained in the following definition.

DEFINITION. *If* M *is a finitely generated* G-*module, where* G *is a positively graded noetherian ring, then the spread* s(M) *of* M *is defined to be the least integer* s *for which there exist homogeneous elements* $x_1,\dots,x_s$ *of* G *of positive degree such that the ideal* $x_1G + \dots + x_sG + \mathrm{Ann}M$ *is irrelevant, or what is equivalent, such that* $(x_1M + \dots + x_sM)_n = M_n$ *for* n *large.*

Note that Theorem 1.12 now implies that, if $G' = G/\mathrm{Ann}M$ and $x_1',\dots,x_s'$ are the images of $x_1,\dots,x_s$ in G', then M can be considered as a finitely generated $G_0'[x_1',\dots,x_s']$-module. It now follows that for any a, $s(a,M) \leq s(M)$.

Our main concern will be with the case when G_0 is local. In this case the basic result is the following.

THEOREM 1.43. *Let* G *be a positively graded noetherian ring,* G_0 *being local with maximal ideal* $\mathfrak{m}$, *and let* M *be a finitely generated graded* G*-module. Then if* $G' = G/\mathfrak{m}G$, *and* M' *is the* G'*-module* $M/\mathfrak{m}M$,

$$s(M) = s(M').$$

The statement that $(x_1M + \ldots + x_sM)_n = M_n$ is, by Nakayama's lemma, equivalent to the statement that

$$(x_1M + \ldots + x_sM)_n + \mathfrak{m}M_n = M_n$$

i.e., to the statement that

$$(x_1'M' + \ldots + x_s'M')_n = M'_n$$

where x_i' is the image of x_i in G'. The equality s(M) = s(M') is now immediate.

The above theorem essentially reduces the study of s(M) in the case where G_0 is local to the case where G_0 is a field k, where we can assume that $a = \dim_k$. In this case, as we shall now prove, s(a,M) = s(M).

THEOREM 1.44. *Let* G *be a positively graded noetherian ring with* G_0 *a field* k. *If* V *is a vector space over* k *let* a(V) *denote its dimension. Then, for any finitely generated graded* G*-module* M,

$$s(M) = s(a,M).$$

First, we note that we can replace G by G/AnnM, and suppose that AnnM = (0). Now we have already remarked that $s(a,M) \leq s(M)$. Hence we have to show that $s(a,M) \geq s(M)$.

We proceed by induction on s(a,M). First suppose that s(a,M) = 0. Then M_n has zero dimension if n is large, i.e., M_n is zero for n large, and hence s(M) = 0.

Now suppose that s(a,M) > 0. This implies that s(M) > 0. Now let N be the graded sub-module of M consisting of all elements annihilated by an irrelevant ideal, i.e., by a power of g_1. Then, as N is finitely generated, $N_n = (0)$ for n large. It follows that s(a,N) = 0 and hence that s(a,M) = s(a,M/N). Further, Ann(M/N) is of the form $0 : (g^+)^r$ for some r and hence must be nilpotent since g^+ is not a minimal prime ideal of G.

Hence, for any elements $x_1,\dots,x_q$ of G,

$$(x_1G + \dots + x_qG) \supseteq (x_1G + \dots + x_qG + \mathrm{Ann}(M/N))^n$$

if n is large, and so it follows that $s(M) = s(M/N)$. Hence we can, by replacing M by M/N, assume that g^+ is not an associated prime of the zero sub-module of M. Then, the corollary to Theorem 1.14, with $I = g^+$ and $\mathbf{p}_1,\dots,\mathbf{p}_s$ taken as the maximal associated prime ideals of (0) in M, implies that there exists a homogeneous element x of positive degree in G which is not a zero divisor on M. Using the arguments at the beginning of section 1.4, and replacing x by a suitable power, if necessary, we can assume that x has degree f where f is such that, for each j, $0 \leq j < f$, $a(M_{rf+j})$ is a polynomial $H_j(r)$ in r for r large, the maximum of the degrees of these polynomials being $s(a,M) - 1$. We now have an exact sequence

$$0 \longrightarrow M(f) \longrightarrow M \longrightarrow M/xM \longrightarrow 0$$

from which it follows that

$$a((M/xM)_n) = a(M_n) - a(M_{n-f})$$

for all large n and hence that

$$a((M/xM)_{rf+j}) = H_j(r) - H_j(r-1)$$

for large r. It follows that $s(a,M/xM) \leq s(a,M)-1$. Now we apply our inductive assumption and find elements $x_2,\dots,x_s$ of G, where $s = s(a,M)$, such that $\mathrm{Ann}(M/xM) + x_2G + \dots + x_sG$ is irrelevant. We now require the following lemma.

LEMMA 1.441. *If* M *is a finitely generated* A-*module and* I *is an ideal of* A, *then* Ann(M/IM) *is contained in the radical of* AnnM + I.

Let $u_1,\dots,u_s$ be a set of generators of M. Then if x belongs to AnnM/IM, we can write

$$xu_i = a_{i1}u_1 + \dots + a_{is}u_s, \quad i = 1,\dots,s$$

where a_{ij} belongs to I. Hence $\det(xd_{ij} - a_{ij})u_i = 0$ for $i = 1,\dots,s$, and hence x^s belongs to I + AnnM, which proves the lemma.

Returning to the proof of Theorem 1.44, it now follows that AnnM/xM is contained in the radical of xG, and hence $(x,x_2,\dots,x_s)$ is an irrelevant ideal of G. Hence $s(M) \leq s$ and this completes the proof.

We continue with a simple and well-known result for the case $G_0 = k$.

THEOREM 1.45. *Let* G *be a positively graded noetherian ring with* $G_0 = k$, *and let* M *be a finitely generated graded* G*-module. Then, if* $\mathbf{p}_1,\dots,\mathbf{p}_s$ *are the minimal prime ideals associated with* M, *and* t_i *denotes the transcendence degree of the field of fractions of* $G/\mathbf{p}_i$ *over* k, $s(M) = \mathrm{Max}(t_i)$.

We consider a filtration of M whose factors are all of the form $(G/\mathbf{p})(n)$, where $\mathbf{p}$ denotes a prime ideal of G. By repeated use of a),b) on p. 21, it follows that $s(M) = s(a,M) = \mathrm{Max}\ s(a,G/\mathbf{p})$, taken over those $\mathbf{p}$ for which $(G/\mathbf{p})(n)$ is a factor of the filtration and, by b) on p. 21 we can restrict $\mathbf{p}$ to the minimal primes associated with M. We have thus reduced the proof to the case where M = G and G a domain, and we have to prove that s(G) = t, where t is the transcendence degree of the field of fractions of G. We now deal with this case.

If s = s(G), then we can find elements $x_1,\dots,x_s$ of positive degree such that G is a finite $k[x_1,\dots,x_s]$-module. Further, we can replace $x_1,\dots,x_s$ by suitable powers and assume that they all have the same degree. Since the field of fractions of G is a finite algebraic extension of $k(x_1,\dots,x_s)$, we can assume that $G = k[x_1,\dots,x_s]$, and further we can assume that $x_1,\dots,x_s$ all have degree 1. We can therefore assume that G is a homomorphic image of $k[X_1,\dots,X_s]$ and that s(a,G) = s. But $a(k[X_1,\dots,X_s]_n) = \binom{n+s-1}{s-1}$ has degree s-1, and if the kernel of the homomorphism of $k[X_1,\dots,X_s]$ onto G is non-zero, then G is a homomorphic image of $k[X_1,\dots,X_s]/(f)$, where f is a non-zero homogeneous polynomial, of degree m say. Then

$$a(G_n) \le \binom{n+s-1}{s-1} - \binom{n-m+s-1}{s-1}.$$

Hence the polynomial representing $a(G_n)$ for large n has degree at most s-2, which is a contradiction. Hence G is isomorphic to $k[X_1,\dots,X_s]$ and so its field of fractions has transcendence degree s over k.

We conclude this chapter with a technical result which will be used in chapter 6 section 1.

DEFINITIONS. *Let* G *be a graded ring and let* M *be a* G*-module. By the* r^{th} *Veronese ring of* G *we mean the graded ring* G<r> *such that*

$$G\langle r\rangle_n = G_{rn}.$$

The set of rth *Veronese modules of* M *consists of the modules* $M\langle r,j\rangle$, $j = 0,\dots,r-1$, *such that*

$$M\langle r,j\rangle_n = M_{rn+j}.$$

THEOREM 1.46. i) *If* G *is a graded noether ring, and* M *is a finitely generated graded* G*-module, then* $G\langle r\rangle$ *is a graded noether ring, and each of the modules* $M\langle r,j\rangle$ *is a finitely generated* $G\langle r\rangle$ *module.*

ii) *If* G *is positively graded and* G_0 *is a field* k, *then*

$$s(M) = \text{Max } s(M\langle r,j\rangle),$$

the maximum being taken over $j = 0,\dots,r-1$. *If* $M = G$, *then* $s(G) = s(G\langle r\rangle)$.

i) By an abuse of notation, let $G\langle r\rangle$ be temporarily identified with the graded ring obtained from $G\langle r\rangle$ by multiplying degrees by r. This will not alter the property of being noetherian or non-noetherian. With this change $G\langle r\rangle$ is a sub-ring of G. Let I be an ideal of $G\langle r\rangle$. Then IG is finitely generated by a finite set of elements of I. Since $IG \cap G\langle R\rangle = I$, these elements generate I as an ideal of $G\langle r\rangle$.

Next we note that G is finitely generated over $G\langle r\rangle$ and the rth power of any element of G belongs to $G\langle r\rangle$. Hence G is a finite $G\langle r\rangle$-module and M, as a finite G-module, is a finite $G\langle r\rangle$-module. The modules $M\langle r,j\rangle$, treated in a similar way to $G\langle r\rangle$, are then $G\langle r\rangle$-sub-modules of M and so are finitely generated.

ii). We use the equality $s(M) = s(a,M)$ of Theorem 1.44. The value of $s(a,M)$ is unaltered if we consider M as a $G\langle r\rangle$-module in place of a G-module. Now we can define $s(a,M)$ as the least integer s such that $n^{-s}a(M_n) \to 0$ as $n \to \infty$. This is equivalent to $(nr+j)^{-s}a(rn+j) \to 0$ as $n \to \infty$ for $j = 0,\dots,r-1$, and finally to $n^{-s}a(nr+j) \to 0$ for $j = 1,\dots,r-1$ as $n \to 0$. Hence $s(a,M) = \text{Max } s(a,M\langle j\rangle)$, $j = 1,\dots,r-1$, as required. If $M = G$, then the modules $G\langle r,j\rangle$ are all $G\langle r\rangle$-modules, and hence $s(G\langle r,j\rangle) \leq s(G\langle r\rangle) = s(G\langle r,0\rangle)$ and the last statement follows.

2. FILTRATIONS AND NOETHER FILTRATIONS

1. Generalities on filtrations.

To commence this section, A will be an arbitrary commutative ring.

DEFINITION. *A filtration on* A *is a function on* A *taking as values real numbers together with the value* ∞, *and satisfying the following conditions*

i) $f(1) \geq 0$; $f(0) = \infty$,

ii) $f(x - y) \geq \mathrm{Min}\,(f(x),f(y))$,

iii) $f(xy) \geq f(x) + f(y)$.

The axioms above permit two degenerate types of filtration, which we allow since they may arise as a result of some of our constructions. Note that iii) implies

$$f(1) \geq f(1) + f(1).$$

First this permits $f(1) = \infty$. If this is the case then, for any x in A, we have, again as a result of iii),

$$f(x) \geq f(x) + f(1),$$

and hence $f(x) = \infty$ for all x in A. This is the first degenerate case. If however $f(1) \neq \infty$, then $f(1) \leq 0$, and the first half of i) implies that $f(1) = 0$. Our second degenerate case is that $f(x) = 0$ for all $x \neq 0$ and $f(0) = \infty$. This is clearly permitted by the axioms.

Although we allow arbitrary real numbers as values of a filtration, we will be concerned mainly with filtrations whose values are integers together with ∞. In fact, given a general filtration f as above, we can always approximate to it by a filtration taking integer values (including ∞). We could take, for example, the filtration $[f](x)$ defined as the integer part $[f(x)]$ of $f(x)$.

An important role will be played below by an equivalence relation between filtrations. We will term two filtrations *f,g equivalent*, and write $f \equiv g$, if, for all x, $|f(x) - g(x)| < K$ for some $K > 0$ and independent of x (this is understood to mean that if either $f(x)$ or $g(x)$ is equal to ∞ then so is the other). It is clear that this is an equivalence relation, and further that $f(x)$ and $[f](x)$ are equivalent.

We now consider some special types of filtration. A filtration h is said to be *homogeneous* if for all non-negative integers n,

$$h(x^n) = nh(x).$$

In fact, it is sufficient to assume that $h(x^2) = 2.h(x)$ for all x. For this implies that, if m is a power of 2, then $h(x^m) = mh(x)$. Let m be a power of 2 greater than n. Then, using iii)

$$m.h(x) = h(x^m) = h(x^n.x^{n-m}) \geq h(x^n) + (m-n)h(x).$$

Hence $n.h(x) \geq h(x^n)$. But by repeated use of iii), $h(x^n) \geq n.h(x)$.

Next we term a filtration v a *valuation* if

$$v(xy) = v(x) + v(y)$$

for all x,y. It is clear that if (v_i) is a family of valuations, then $h(x) = \text{Inf} v_i(x)$ is a homogeneous filtration. We will be concerned particularly with the case when the family is finite, when we shall term $\text{Inf} v_i(x)$ a *sub-valuation.*

We now prove two general lemmas on filtrations.

LEMMA 2.11. *If* $f(x)$ *is a filtration on* A, *then the limit of* $f(x^n)/n$ *as* n *tends to infinity for all* x *in* A *exists, if* ∞ *is allowed as a value for the limit. Further, if we denote this limit by* $\mathfrak{f}(x)$, *then* $\mathfrak{f}(x)$ *is a homogeneous filtration, and, if* $h(x)$ *is any homogeneous filtration satisfying* $h(x) \geq f(x)$ *for all* x *in* A, *we have also* $h(x) \geq \mathfrak{f}(x)$ *for all* x *in* A .

If $f(x^m) = \infty$ for some positive integer m, then $f(x^n) = \infty$ for all $n > m$. Hence in this case the limit is equal to ∞. We may now assume that $f(x^n) < \infty$ for all n. Write $a(n)$ for $f(x^n)/n$. Then

$$(m + n)a(m + n) = f(x^{m+n}) \geq f(x^m) + f(x^n) = ma(m) + na(n),$$

Now let $a = \text{Sup} a(n)$ and let $b < a$. Then we can find $m > 0$ such that $a(m) > b$. If $n>m$, we can write $n = qm + r$ with $0 \leq r < m$. Then since $a(qm) \geq a(m) > b$,

$$a(n) \geq (qma(mq) + ra(r))/n > (qma(m) + ra(r))/n$$
$$= a(m) + (r(a(r) - a(m))/n).$$

Now the set of numbers $r(a(r) - a(m))$ are finite in number and all finite. Hence, if n is sufficiently large, $a(n) > b$. Hence $a(n) = f(x^n)/n$ tends to the limit a as n tends to infinity.

Next write $a(n) = f(x^n)/n$ as above, and $b(n) = f(y^n)/n$, $c(n) = f((x-y)^n)/n$. Then

$$\mathfrak{f}(x-y) \geq c(n) \geq \text{Min}(f(x^n),f(x^{n-1}y),\ldots,f(y^n))/n \geq (sa(s) + (n-s)b(n-s))/n$$

for some s between 0 and n.

Now let $a = \text{Min}(\mathfrak{f}(x),\mathfrak{f}(y))$ and let $b < a$. Then we can find m such that if $n > m$, then $a(n),b(n) > b$. Let q be any integer >1, and put $n = qm$. Then

$$c(qm) \geq sa(s) + (qm - s)b(qm - s)/qm \quad \text{for some } s \leq qm.$$

If both s and $(qm - s)$ are $\geq m$, this implies that $c(mq) > b$. If not, then $c(mq) > (q-1)b/q$. Hence $\mathfrak{f}(x-y) > (q-1)b/q$ for all q, and this implies that $\mathfrak{f}(x-y) \geq b$ and hence $\mathfrak{f}(x-y) \geq \text{Min}(\mathfrak{f}(x),\mathfrak{f}(y))$.

Next $f(x^n y^n) \geq f(x^n) + f(y^n)$ implies that $\mathfrak{f}(xy) \geq \mathfrak{f}(x) + \mathfrak{f}(y)$. The truth of condition i) for $\mathfrak{f}$ is immediate. Hence $\mathfrak{f}$ is a filtration on A.

That it is homogeneous follows since, if $y = x^m$, and $a(n),b(n)$ are as above, $b(n) = ma(nm)$, and letting n tend to infinity, we see that

$$\mathfrak{f}(x^m) = m\mathfrak{f}(x).$$

Finally, if $h(x)$ is a homogeneous filtration such that $h(x) \geq f(x)$ for all x, then $h(x) = h(x^n)/n \geq f(x^n)/n = a(n)$. Letting n tend to infinity we obtain the required result that $h(x) \geq \mathfrak{f}(x)$.

COROLLARY i). $\mathfrak{f}(x) \geq f(x)$ *for all* x.

The proof is immediate.

COROLLARY ii). If $f \equiv g$, *then* $\mathfrak{f} = \mathfrak{g}$.

The proof is immediate.

LEMMA 2.12. *Let* $h(x) = \text{Min}(v_1(x),\ldots,v_s(x))$, *where* $v_1(x),\ldots,v_s(x)$ *are valuations on* A. *Further, suppose that this representation is irredundant. Then* $v_1,\ldots v_s$ *are uniquely determined by* h.

Suppose that F is the sub-set of A consisting of all elements x such that $h(x) < \infty$. Then $v_i(x) = \infty$ for all x not in F.

Now suppose that $v(x)$ is any valuation on A such that $v(x) \geq h(x)$ for all x in A. Let S_v denote the sub-set of F consisting of all elements x such that $v(x) = h(x)$. Then, if $x_1,\ldots,x_r$ belong to S_v,

$$h(x_1x_2\ldots x_r) \leq v(x_1\ldots x_r) = v(x_1) + \ldots + v(x_r) = h(x_1) + \ldots + h(x_r) \leq h(x_1x_2\ldots x_r)$$

whence

$$h(x_1\ldots x_r) = h(x_1) + \ldots + h(x_r).$$

A sub-set S of F with the property that this equation holds for any finite set of elements $x_1,\ldots,x_r$ of S will be termed *h-compatible*. Hence S_v is h-compatible for any valuation v on A such that $v(x) \geq h(x)$ for all x in A.

We now observe that, as h is homogeneous, any sub-set of F containing one element is h-compatible. Hence h-compatible sub-sets exist, and further we can apply Zorn's Lemma to see that F is the union of its maximal h-compatible sub-sets.

We now show that the maximal h-compatible sub-sets of F are the sets S_v for $v = v_1,\dots,v_s$. We write these sets as $S_1,\dots,S_s$. Suppose that S is an h-compatible sub-set of F not contained in any of the sub-sets $S_1,\dots,S_s$. Then, for each i, there is an element x_i of S not in S_i, i.e. $v_i(x_i) > h(x_i)$. Hence

$$h(x_1) + \dots + h(x_s) < v_i(x_1) + \dots + v_i(x_s) = v_i(x_1\dots x_s)$$

for each i, and

$$h(x_1) + \dots + h(x_s) < \mathrm{Min}(v_1(x_1\dots x_s),\dots,v_s(x_1\dots x_s)) = h(x_1\dots x_s)$$

which contradicts S being h-compatible. Hence $S_1,\dots,S_s$ are h-compatible sub-sets of F and a maximal h-compatible sub-set is one of the sets S_i. Next $S_1,\dots,S_s$ are distinct: for the condition of irredundancy states that for each i there is an element x_i in A such that $h(x_i) = v_i(x_i) < v_j(x_i)$ for all $j \neq i$. This implies that $h(x_i) < \infty$, and hence x_i is contained in F. Further, S_i is not contained in the union of the sets S_j for $j \neq i$. Hence the sets S_i are the maximal h-compatible sub-sets of F.

Now define Σ_i to be the set of elements a in F such that $aS_i \cap S_i \neq \emptyset$. Then Σ_i consists of all a such that $v_i(a) < \infty$. For suppose a is such an element. Then, if x_i is as above and $j \neq i$,

$$v_j(ax_i^m) - v_i(ax_i^m) = v_j(a) - v_i(a) + m(v_j(x_i) - v_i(x_i)) > 0$$

if m is large enough. Hence $aS_i \cap S_i$ contains ax_i^m and so a belongs to Σ_i.

Conversely, if $ay = z$, with y,z in S_i,

$$v_i(a) = v_i(z) - v_i(y) = h(z) - h(y) < \infty.$$

This shows that v_i is determined on Σ_i by h and is ∞ outside Σ_i. Hence the valuations $v_1,\dots,v_k$ are uniquely determined by h.

2. Integer-valued filtrations.

In this section we impose the restriction that the filtration f under consideration takes on values which are integers or ∞. The same will not be true of some of the filtrations derived from f, particularly $\mathfrak{f}$.

We therefore consider an integer-valued filtration f on a commutative ring A. We first consider the set of elements of A which satisfy the condition that $f(x) \geq 0$. These elements form a sub-ring of A which we will denote by $A_0(f)$ or simply A_0. Next suppose that n is a positive integer. Then the set of elements of A such that $f(x) \geq n$ is an ideal of A_0 which we denote by $I_n(f)$, or simply I_n if there is no danger of confusion. Finally, if n is a negative integer, then the set of elements of A such that $f(x) \geq n$ is an A_0-module which we also denote by $I_n(f)$ or I_n. Note that A is the union of the A_0-modules $I_n(f)$.

Note that we have, for all m,n,

$$I_{m+n} \supseteq I_m . I_n$$

so that the ideals I_n, $n > 0$, all have the same radical which we term the *radical* of f.

It follows from the above that the restriction of f to A_0 is a filtration on that ring which takes only non-negative values, and which we will denote by f^+; f also determines a filtration on A taking only non-positive values, and which we denote by f^- and define, if $x \neq 0$, by

$$f^-(x) = \text{Min}\{f(x),0\}.$$

The two filtrations $f^+(x)$, $f^-(x)$ together define f.

Our main concern, however, is with a graded ring G(f) associated with f.

DEFINITION. *The graded ring* G(f) *of* f *is defined to be the sub-ring of* $A[t,t^{-1}]$, *where* t *is an indeterminate over* A, *consisting of all finite sums* $a_{-p}t^{-p}+\dots+a_qt^q$ *where* $f(a_j) \geq j$.

It will be convenient to denote t^{-1} by u in what follows.

First we note that G(f) is a sub-ring of A[t,u] which contains u, and in fact the sub-ring $A_0[u]$.

It is convenient at this point to consider a general example of a filtration on a ring A. Suppose that I is an ideal of a ring A_0 contained in A and such that for all elements x of A, $A_0 \supseteq I^n.x$ for some n. Then we can define a filtration f_I on A as follows. If x belongs to A_0, then $f_I(x)$ is the largest integer such that x belongs to I^n, if such an integer is defined, and $f_I(x) = \infty$ if x belongs to all powers of I. If x does not belong to A_0, then $f_I(x) = -n$, where n is the smallest integer such that $A_0 \supseteq I^n.x$. It is easy to verify that f_I as defined is a filtration on A. We are particularly concerned at present with the case where I is a principal ideal aA_0 of A_0 generated by a non-zero divisor of A_0 and A is the ring of fractions $(A_0)_{(a)}$, where (a) denotes the set of powers of a. We will term this filtration a *principal filtration* on A and denote it by f_a.

Now suppose that f is an arbitrary integer-valued filtration on a ring A. We consider the ring A[t,u] and its sub-ring G(f). We have seen that u is contained in G(f), and is clearly a non-zero divisor of G(f). Further, $A[t,u] = G(f)_{(u)}$. Hence we can use the above construction to define a filtration f_u on A[t,u]. Now A is a sub-ring of A[t,u] consisting of the elements of degree 0. Consider the restriction of f_u to A. First suppose x belongs to A_0. Then x belongs to G(f) and hence $f_u(x)$ is the largest integer n such that x belongs to $u^nG(f)$ or, what is the same thing, the largest integer n such that xt^n belongs to G(f), if such a largest integer is defined, and is ∞ otherwise. Hence if x belongs to A_0, $f_u(x) = f(x)$. Now suppose x does not belong to A_0, i.e., x does not belong to G(f). Then in this case $f_u(x) = -n$ where n is the least integer such that u^nx belongs to G(f), and so again is equal to f(x). It follows that f is the restriction of the principal filtration f_u to A. This enables us to prove a number of results on filtrations by restriction to principal filtrations. Note that the above construction has an inverse. Suppose G is a graded sub-ring of A[t,u] containing u and with the property that $G_{(u)} = A[t,u]$. Then the restriction of f_{uG} to A defines a filtration f on A, such that G(f) = G. Hence filtrations on A correspond to graded sub-rings of A[t,u] which are "large enough" in the sense that they contain u and, for all x in A, xu^n for n sufficiently large. In particular, if f is a filtration on A and G' is a graded sub-ring of A[t,u] containing G(f), then G' determines a filtration f' on A which satisfies $f'(x) \geq f(x)$ for all x.

As already observed, the filtration $\mathfrak{f}(x)$ is not necessarily integer-valued when f(x) is. Our next aim is to define a filtration f*(x) which is integer-valued whenever f(x) is, and which is equivalent to $\mathfrak{f}(x)$. f* will be termed the *integral closure* of f.

In fact, at the level of generality of this section, there are two candidates for f*(x). The one we shall adopt is the following. We define $f^*(x) \geq m$ to mean that x satisfies an equation

$$x^n + a_1x^{n-1} + \dots + a_n = 0$$

where $f(a_i) \geq mi$. This is equivalent to the element xt^m being integrally dependent on G(f). However, it is not a simple matter to prove that the integral closure of G(f) in A[t,u] is a graded sub-ring of A[t,u]. An elegant proof of a more general result implying this statement appears in [BAC], Chapter V, paragraph 1.8 as Proposition 20. Assuming this fact, the integral closure of G(f) in A[t,u], which we will denote by G*(f), determines, as above, a filtration on A which is f*(x), and G*(f) = G(f*).

This difficulty does not arise if we replace the integral closure of G(f) by the complete integral closure of G(f). (Recall that, whereas an element x of A[t,u] belongs to the integral closure of G(f) if the positive powers of x generate a finite G(f)-module, x belongs to the complete integral closure of G(f) if the powers of x belong to a finitely generated G(f)-module.) This is clearly equivalent to the condition that there exists an integer k such that x^n belongs to $u^{-k}G(f)$ for all positive n. We denote the complete integral closure of G(f) in A[t,u] by G'(f). Then G'(f) is a graded ring. What we have to show is that, if $x = \Sigma x_it^i$ belongs to G'(f), then so do the elements x_it^i. We prove this by induction on the number of values of i for which x_i is non-zero. If this is 1, the result is obvious. If it is greater than 1, let r be the greatest value of i for which x_i is not zero. Then, if x^n belongs to the graded G(f)-module $u^{-k}G(f)$ for all positive n, it is clear that $x_r{}^nt^{rn}$ belongs to $u^{-k}G(f)$ for all n and hence x_rt^r belongs to G'(f). Subtracting this term from x, the remaining terms x_it^i also belong to G'(f) by our inductive hypothesis.

Finally we note that if, as in the next section, G(f) is noetherian, the notions of integral closure and complete integral closure coincide, and the two approaches are equivalent.

LEMMA 2.21. *If f is a filtration on A, then*

$$f(x) \le f^*(x) \le \mathfrak{f}(x) \le f^*(x) + 1.$$

The first inequality follows since $G^*(f) \supseteq \bar{G}(f)$.

To prove the next inequality, suppose that $f^*(x) \ge m$. Then x satisfies an equation

$$x^r + a_1 x^{r-1} + \dots + a_r = 0$$

where $f(a_i) \ge mi$. We now prove by induction on n that x satisfies an equation

$$x^{n+r} + a_{1n} x^{r-1} + \dots + a_{rn} = 0,$$

where $f(a_{in}) \ge (n+i)m$. Suppose this proved for n. Then

$$x^{n+r+1} = -a_{1n}x^r - \dots - a_{rn}x = +a_{1n}(a_1 x^{r-1} + \dots + a_r) - \dots - a_{rn}x.$$

The coefficient of x^{r-i} is $a_{1n}a_i - a_{i+1,n}$ and the result follows since $f(a_{1n}a_i) \ge (n+1)m + im = (n+1+i)m$, and $f(a_{i+1,n}) \ge (n+i+1)m$. Hence we have

$$f(x^{n+r}) \ge \text{Min}\, f(-a_{in}x^{r-i}) \ge \text{Min}\,((n+i)m + f(x^{r-i}))$$

from which it follows that $\mathfrak{f}(x) \ge m$.

To prove the last inequality, suppose that $f^*(x^n) \ge n(m+1)$. Then x satisfies an equation

$$x^{nr} + b_1 x^{n(r-1)} + \dots + b_r = 0,$$

where $f(b_i) \ge \text{Min}(m+1)$. It then follows that $f^*(x) \ge m+1$. Hence, if $f^*(x) = m < \infty$, $f(x^n) \le f^*(x^n) < n(m+1)$. It follows that $f(x^n)/n < m+1$ for all n, proving that $\mathfrak{f}(x) \le m+1$. The case $f^*(x) = \infty$ is immediate, if we interpret $\infty+1$ as ∞.

It follows that $f^*(x)$ is equivalent to $\mathfrak{f}(x)$, and we will often use it in place of $\mathfrak{f}(x)$.

3. Noether filtrations.

We commence this section with a definition.

DEFINITION. *A filtration f on a commutative ring A is termed a noether filtration if the associated graded ring G(f) is a graded noetherian ring.*

We now consider some of the consequences of Samuel's Theorem for this definition. First $G_0(f)$ is a noetherian ring, i.e., the ring A_0 is noetherian. Next the rings $G^+(f)$ consisting of the elements of non-negative degree, and $G^-(f)$ consisting of the terms of non-positive degree, are both noetherian graded rings. Now consider the filtration $f^+(x)$ on A_0. Since all elements x of A_0 satisfy $f^+(x) \geq 0 > -n$ for any positive integer n, it follows that $G(f^+)$ is not $G^+(f)$, but $G^+(f)[u]$. However, $G^+(f)[u]$ is a graded noetherian ring if and only if $G^+(f)$ and $A_0[u]$ are noetherian by Samuel's Theorem. The latter is noetherian since A_0 is noetherian. This implies that $f^+(x)$ is a noether filtration. Next we turn to $f^-(x)$. The associated graded ring of this filtration is $G^-(f)$ and consequently this filtration is also noetherian.

We delay further consideration of the consequences of Samuel's Theorem for noether filtrations to the end of this section and instead continue with a number of other results which indicate the simplifying nature of the noetherian assumption. First we consider the integral closure f^* of a noether filtration f. Recall that f^* is determined by the integral closure of the ring G(f) in A[t,u]. If we assume that G(f) is noetherian, then xt^m belongs to the integral closure of G(f) if and only if there exists an integer k such that $(xt^m)^n \in u^{-k}G(f)$ for all n. Hence we have

LEMMA 2.31. *If* f *is a noether filtration, then* $f^*(x) \geq m$ *if and only if there exists a constant* k *(depending on x) such that* $f(x^n) \geq mn-k$ *for all* $n > 0$.

Next suppose that f is a filtration on A, and that $\mathfrak{a}$ is an ideal of A. Then we can define on the quotient ring $A/\mathfrak{a}$ a filtration denoted by $f/\mathfrak{a}$ by defining $I_n(f/\mathfrak{a}) = (I_n+\mathfrak{a})/\mathfrak{a}$. It is clear that $f/\mathfrak{a}$ is noetherian, since $G(f/\mathfrak{a})$ is the homomorphic image of G(f) under the restriction of the natural map $A[t,u] \rightarrow (A/\mathfrak{a})[t,u]$.

We make use of this construction in reducing the discussion of f^* largely to the case where A is an integral domain.

LEMMA 2.32. *Let* $\mathfrak{p}_1,\ldots,\mathfrak{p}_s$ *be the minimal prime ideals of a noether ring* A *and let* f *be a noether filtration on* A. *Then*

$$f^*(x) = \text{Min}\,(f/\mathfrak{p}_i)^*(x_i),$$

the minimum being for $i = 1,\ldots,s$, *and* x_i *denoting the image of* x *in* $A/\mathfrak{p}_i$.

Write $f_i(x)$ for $(f/\mathfrak{p}_i)(x)$. Then it is clear that $f_i(x_i) \geq f(x)$ for all i, whence $f(x) \leq \text{Min}\, f_i(x_i)$. On the other hand, if $f_i(x_i) \geq m$ for all i, we can find a_i in $I_m(f)$ for each i such that $x-a_i \in \mathfrak{p}_i$. But there exists an integer k such that $(\mathfrak{p}_1\mathfrak{p}_2\ldots\mathfrak{p}_s)^k = 0$. Hence $((x-a_1)\ldots(x-a_s))^k = 0$, and this implies that $f^*(x) \geq m$. Hence $\text{Min}(f_i(x_i)) \leq f^*(x)$. Hence we have

$$f^*(x) \leq \text{Min } f_i^*(x_i) \leq f^*(x)$$

which proves the result.

The two parts of the following theorem are, respectively, variants of the Krull Intersection Theorem and the Artin-Rees Lemma.

THEOREM 2.33. i). *Let* $f(x)$ *be a noether filtration on* A. *Then there exists an element* a *in* A *satisfying* $f(a) > 0$ *such that* $f(x) = \infty$ *if and only if* $ax = x$.

ii). *If* x *is not a zero divisor on* A, *and* f *is a noether filtration on* A, *such that* $f(1) \neq \infty$ *then there exists an integer* $\mu(x)$, *depending only on* x *and* f, *such that if* $f(y) < \infty$,

$$f(xy) \leq f(y) + \mu(x).$$

i) We first consider a single element x and show that $f(x) = \infty$ if and only if there exists an element $a(x)$ with $f(a(x)) > 0$ such that $a(x)x = x$. The "if" part is obvious. To prove "only if", consider the filtration f_u on $G(f)$. Then $f_u(x) = \infty$, i.e., x belongs to $u^nG(f)$ for all n. Hence the elements t^nx belong to $G(f)$ for all n. Hence, if they generate an ideal J, there exists an integer m such that J is generated by the elements t^rx, $1 \leq r \leq m$. It follows that

$$t^{m+1}x = \Sigma\, a_rt^r.xt^{m+1-r},$$

summed from r = 1 to m, where a_rt^{m+1-r} belongs to $G(f)$, i.e., $f(a_r) \geq m+1-r \geq 1$. Then if $a(x) = \Sigma a_r$, $x = a(x)x$ and $f(a(x)) \geq 1 > 0$.

Now let N be the ideal of A consisting of all elements x such that $f(x) = \infty$. Let $x_1,\ldots,x_m$ be a basis of N and let $a(i) = a(x_i)$ as defined above. Then, for all x in N,

$$((1-a(1))\ldots(1-a(m)))x = 0.$$

Now we can write $(1-a(1))\ldots(1-a(m))$ in the form $1-a$ and clearly $f(a) > 0$. We now have $x = ax$ for all x such that $f(x) = \infty$ as required. Note that this implies that, if x is not a zero divisor, then $f(x) < \infty$.

ii) As above, we replace A by $G(f)$ and f by the principal filtration f_u whose restriction to A is f. Now suppose that $f_u(xy) \geq n$. This is equivalent to $xu^{-n}y$ belonging to $G(f)$. But this, in turn, is equivalent to $u^{-n}y$ belonging to $(G(f): x)_{A[t,u]}$. Since x is not a zero divisor, this is isomorphic as a G(f)-module to the ideal $x(G(f):x)_{A[t,u]}$ of G(f),and, as G(f) is noetherian, is a finite G(f)-module contained in $A[t,u] = G(f)_{(u)}$. It follows that there exists an integer μ such that

$$u^{-\mu}G(f) \supseteq (G(f):x)_{A[t,u]}.$$

Then $u^{\mu-n}y$ belongs to G(f) and hence $f_u(y) \geq n - \mu$. As $f(x), f(y) < \infty$, this implies that $f(xy) < \infty$, and we can take $f_u(xy) = n$. Hence $f(xy) - f(y) \leq \mu$ for all y such that $f(y) < \infty$. We take $\mu(x) = \mu$.

Note that the above shows that if $f(xy) = \infty$, then $f(y) = \infty$.

COROLLARY i). *If x is not a zero divisor, and* $f(1) = 0$, *then* $f^*(x) < \infty$.

First, $f(x^n) < \infty$ for all n by 2.33 i). Then, if $\mu(x)$ is as above, it follows that $f(x^n) \leq n\mu(x)$, and hence $f^*(x) \leq \mathbf{f}(x) < \mu(x) < \infty$.

COROLLARY ii). *Let f be a noether filtration on A such that* $f(1) < \infty$, *and let J be the ideal of A consisting of all y such that* $f(y) = \infty$. *Then* $f^*(x) = \infty$ *if and only if x belongs to the radical of J.*

If x does belong to the radical of J, then $f(x^r) = \infty$ for some r, and consequently $\mathbf{f}(x) = \infty$ and hence $f^*(x) = \infty$. To prove the converse, we first note that, by replacing A,f by A/J,f/J, we can assume that J = 0, and this restriction is in force below. Now if $f^*(x) = \infty$, x is a zero divisor by Corollary i). Hence the result is true if A is a domain, since then $f^*(x) = \infty$ implies $x = 0$. In general, suppose that A has minimal prime ideals $\mathbf{p}_1,\dots,\mathbf{p}_s$. Then as, by Lemma 2.32,

$$f^*(x) = \mathrm{Min}((f/\mathbf{p}_i)^*(x) \mid i = 1,\dots,s),$$

the result will follow from the domain case providing that the condition $(f/\mathbf{p}_i)(1) < \infty$ holds for each i. Suppose that $(f/\mathbf{p}_i)(1) > 0$. Then there exists b in $\mathbf{p}_i$ such that $f(1-b) > 0$. Now suppose that A is not a domain. Then $(0:\mathbf{p}_1) \neq (0)$ and so contains a non-zero element y. Then, by = 0, i.e., $(1-b)y = y$ and hence $f(y) = \infty$, contradicting the assumption that $J = (0)$. Hence $(f/\mathbf{p}_i)(1) = 0$ for each i, and the result is proved.

Next we consider a result and an example which indicates a difference between noether filtrations taking only values ≥ 0 and those which take some negative values.

LEMMA 2.34. *Let* f,g *be two noether filtrations on a noetherian ring* A. *Then*

i) *if* f,g *are equivalent,* $f^*(x) = g^*(x)$ *for all* x *in* A;

ii) *if* f,g *satisfy the two conditions,*

a) $f^*(x) = g^*(x)$ *for all* x *in* A,

b) $A_0(f) = A_0(g)$,

then f,g *are equivalent.*

i) If f,g are equivalent, then there exists a constant C such that $|f(x)-g(x)| < C$ for all x such that $f(x) \neq \infty$, and further $g(x) = \infty$ if and only if $f(x) = \infty$. We now consider the sub-rings G(f),G(g) of A[t,u]. We have to show that the integral closures G*(f), G*(g) of these rings in A[t,u] are equal. Suppose therefore that xt^m belongs to G*(f). Then there exists a constant k such that $f(x^n) \geq nm+k$ for all n. It follows that $g(x^n) \geq nm+k-C$ for all n and hence xt^m belongs to G*(g). Hence $G^*(g) \supseteq G^*(f)$ and the result follows by symmetry.

ii) The assumption that $f^*(x) = g^*(x)$ for all x implies that $G^*(f) = G^*(g)$ and hence that every element of G(g) is integrally dependent on G(f). But G(g) is finitely generated over $A_0(g) = A_0(f)$ and hence is contained in a finitely generated integral extension of G(f) contained in A[t,u]. Hence there exists a positive integer C such that G(g) is contained in $u^{-C}G(f)$. This implies that $g(x) \geq f(x) - C$ for all x. We now complete the argument by interchanging the roles of f,g.

COROLLARY. *If* f,g *are noether filtrations on* A *which do not take negative values, then* f,g *are equivalent if and only if* $f^*(x) = g^*(x)$ *for all* x *in* A.

For the restriction on f,g implies that $A_0(f) = A = A_0(g)$.

Some such condition as ii) b) is required in general for $f^* = g^*$ to imply the equivalence of f and g as the following example shows.

We take as $A_0(f)$ a noetherian local domain A_0 of dimension 1 with field of fractions K, A_0 being subject to the two conditions:

i) the integral closure A_0^* of A_0 in K is not a finite A_0-module,

ii) A_0^* is a discrete valuation ring.

Now let u be a non-zero non-unit of A_0, so that $K = A_0[u^{-1}] = A_0^*[u^{-1}]$ since A_0 has dimension 1. We now take f to be the principal filtration defined on K by the sub-ring A_0 and the element u, and g to be the principal filtration defined by A_0^* and u. Clearly $g = f^*$ so that $f^* = g^*$. However i) implies that, for any positive integer k, there exists x_k in A_0^* but not in $u^{-k}A_0$. Then $f(x_k) < -k$ and $g(x_k) \geq 0$, proving that f,g are not equivalent.

There remains the question as to whether such a ring A_0 exists. An example has been given by Nagata[LR] ((E3.2) on p. 206). Since we require this example later, we will describe it briefly, referring to [LR] for more details. We take K to be a field of characteristic $p > 0$ such that, if K^p denotes the sub-field of K consisting of all p^{th} powers, then $[K : K^p]$ is infinite. For example, we could take K to be the field $k(z_1,z_2,\ldots,)$ obtained by adjoining a countable set of indeterminates $z_1,z_2,\ldots$ to an arbitrary field k of characteristic p. We now take $R^\wedge$ to be the complete regular local ring $K[[X_1,\ldots,X_n]]$, where $X_1,\ldots,X_n$ are independent indeterminates over K, and we take R to be the sub-ring of $R^\wedge$ consisting of all formal power series f in $X_1,\ldots,X_n$ over K whose coefficients generate a finite extension over K^p. We refer to [LR] p. 206 for the proof that R is a regular local ring with completion $R^\wedge$.

We now take $n = 1$ and we take c to be $\Sigma z_i X_1{}^i$, where $z_1,z_2,\ldots$ are p-independent elements of K. Then $L = R[c]$ is the required example.

To see this, we follow Nagata's argument. Let $d = c^p$, so that d belongs to R. Now L is isomorphic to $R[X]/(X^p-d)R[X]$ and its completion $L^\wedge$ is isomorphic to $R^\wedge[X]/(X^p-d)R^\wedge[X]$ in which the image of X-c is a nilpotent element. It follows that L is a 1-dimensional local domain whose completion contains an element w such that $w^p = 0$. Now consider the integral closure L^* of L in its field of fractions. Suppose that L^* is a finite L-module. Then if u is any non-zero non-unit of L, there exists an integer k such that $u^{-k}L$ contains L^*. Now consider the principal filtration f_u restricted to L. It follows that, if $f_u^*(x) \geq n+k$, then $f_u(x) \geq n$. Now consider a Cauchy sequence of elements $y_1,y_2,\ldots$ of elements of L whose limit in $L^\wedge$ is the non-zero element w. This sequence is not a null sequence, since $w \neq 0$, and hence there exists an integer n such that $f_u(y_i) = n$ for all large i. On the other hand, the sequence (y_i^p) is a null sequence, and hence, for large i, $f_u(y_i^p) \geq p(n+k+1)$, i.e. $y_i^p - u^{p(n+k+1)}a_i = 0$ with a_i in L. This implies that $f_u{*}(y_i) \geq n+k+1$ and hence that $f_u(y_i) > n$ which is a contradiction. Hence L* is not a finite L-module. We can now take $A_0 = L$.

Finally we return to the further consideration of the consequences of Samuel's theorem for noether filtrations.

We suppose that f is a noether filtration on A satisfying the condition that $f(1) = 0$. Let f^+, f^- have the same meanings as earlier, so that $G(f^+) = G^+(f)[u]$, and $G(f^-) = G^-(f)$, these being noetherian. Now, by the corollary to Samuel's theorem, if $x_1, \ldots, x_r$ is a set of homogeneous generators of the ideal g^+ of $G^+(f)$, then

$$G(f^+) = A_0[x_1, \ldots, x_r, u].$$

We can write the element x_i in the form $a_i t^{w(i)}$ where a_i is an element of A_0 and $w(i)$ is a positive integer satisfying $w(i) \leq f(a_i)$. Now suppose that, for some i, $f(a_i) = \infty$. Let a be an element of A satisfying the conditions that $f(a) > 0$ and $f(x) = \infty$ if and only if $x = ax$ (see Theorem 2.33 i)). Then at belongs to g^+, and

$$a_i t^{w(i)} = a_i (at)^{w(i)}.$$

Hence it follows that we can replace all such generators of g^+ by the single element at. Hence, changing notation slightly, we can assume that $x_i = a_i t^{w(i)}$ with $w(i) \leq f(a_i) < \infty$, for $i = 1, \ldots, r-1$. This statement may be true of x_r. If it is not, then we have $a_r = a$, where a is as above, $w(r) = 1$ and $f(a) = \infty$. This implies that $a^2 = a$, and hence A splits up into a direct sum $A_1 \oplus A_2$ with $A_2 = aA = aA_0$. Now suppose that A contains an element y such that $0 < f(y) < \infty$. If we write $y = y_1 + y_2$ with y_i in A_i, it is clear that $f(y_1) = f(y)$, and hence $0 < f(y_1 + a) = f(y) < \infty$. Hence if $a' = a + y_1$, we have $a'a = a$ and, if $f(x) = \infty$, $a'x = a'(ax) = ax = x$. It follows that we can replace $a_r t = at$ by $a't$ and we have $f(a_r) < \infty$. We are left with the case that A contains no elements x with $0 < f(x) < \infty$. In this case r must be 1 and the single generator of g^+ is at, where a is an idempotent such that $f(x) = \infty$ if and only if $x \in aA$. In this case $G(f^+) = A_0[at, u]$.

We are thus left with the case where the generators $x_i = a_i t^{w(i)}$ all satisfy the condition that $0 < w(i) < f(a_i) < \infty$. We will now drop the condition that $x_1,\dots,x_r$ generate the ideal g^+ of $G^+(f)$ and replace it by the weaker condition that

$$G(f^+) = A_0[x_1,\dots,x_r,u].$$

Suppose that $x_i = a_i t^{w(i)}$ with $0 < w(i) \le f(a_i) < \infty$. Let $n = f(a_i) - w(i)$ and write $x_i' = a_i t^{w(i)+n}$. Then as $x_i = u^n x_i'$, it follows that $G(f^+) = A_0[x_1',\dots,x_r',u]$. Hence we can suppose that the generators x_i satisfy the condition that $f(a_i) = w(i)$ for each i providing that f(x) takes at least one positive finite value.

As already indicated, we have weakened the condition on $x_1,\dots,x_r$ to obtain the above result, in that we do not require that $x_1,\dots,x_r$ generate the ideal g^+ of $G^+(f)$ containing all elements of positive degree. However, it is still true that the ideal of $G^+(f)$ generated by $x_1,\dots,x_r$ contains all elements of sufficiently high positive degree. For clearly g^+ is generated by the elements $u^n x_i$, $n = 0,\dots,w(i)-1$; $i = 1,\dots r$. Choose N so that $N(w(i)-n) > w(i)$. Then

$$(u^n x_i)^N = a_i^N t^{N(w(i)-n)} = a_i t^{w(i)}.a_i^{N-1} t^{(N-1)w(i) - Nn}$$

Then the exponent of t is positive and

$$f(a_i^{N-1}) \ge (N-1)f(a_i) = (N-1)w(i) \ge (N-1)w(i) - Nn$$

so that $(u^n x_i)^N$ belongs to the ideal generated by $x_1,\dots,x_r$. Hence the ideal of $G^+(f)$ generated by $x_1,\dots,x_r$ contains a power of g^+ and so contains all elements of sufficiently high positive degree.

DEFINITION. *Let f be a noether filtration on A. Then a set of elements* $a_1,\dots,a_r$ *of* A_0 *is termed a set of generators of* f^+ *if*

$$G(f^+) = A_0[a_1 t^{w(1)},\dots,a_r t^{w(r)},u], \quad 0 < w(i) \le f(a_i).$$

We term w(i) *the weight of* a_i.

We term it a standard set of generators if either,

a) $0 < w(i) = f(a_i) < \infty$ for $i = 1,\dots,r$, *or*

b) $r = 1$, *and* $G(f^+) = A_0[a_1 t,u]$

where $a_1^2 = a_1$ *and* $f(a_1) = \infty$. *We say that* a_1 *has weight* ∞ *(rather than 1 as above).*

We note for future reference that, if there exists a set of generators of f^+ containing r elements, then the argument preceding the above definition shows that there exists a standard set of generators with at most r elements.

THEOREM 2.35. *Let* $a_1,\ldots,a_r$ *be a set of generators of* f^+, *where* f *is a noether filtration on a noether ring* A. *Then the ideal* $I_n(f)$ *consisting of all elements of* A_0 *such that* $f(x) \geq n \geq 0$ *is generated by those products of the elements* a_i *having weight* $\geq n$, *the weight of a product of elements* a_i *being the sum of the weights of its factors, and* a_i *having weight* $w(i)$.

Since $G(f^+) = A_0[a_1t,\ldots,a_rt,u]$, it follows that a homogeneous element xt^n, $n > 0$, of G(f) can be written as a linear combination over A_0 of products of the elements $a_it^{w(i)}$ of weight $\geq n$. This implies that x is in the ideal generated by products of the elements a_i having weight $\geq n$. But xt^n is in G(f) if and only if $x \in I_n(f)$.

Now we turn to the filtration $f^-(x)$. Here we have $G(f^-) = G^-(f)$, and, again invoking Samuel's theorem, we see that $G(f^-) = A_0[y_1,\ldots,y_s,u]$, where $y_j = b_jt^{-w(j)}$, with $f(b_j) < 0$ and $w(j) \geq -f(b_j)$. The elements $b_1,\ldots,b_s$ belong to $A-A_0$ and satisfy the condition that $A= A_0[b_1,\ldots,b_s,]$. Give b_j the weight $w(j) > 0$. Define the weights of monomials in $b_1,\ldots,b_s$ in the same way as above. Further, if $w(j) = -f(b_j) + n$, and $n > 0$, we can always replace the generators y_j by $y_j' = b_jt^{w(j)+n}$.

DEFINITION. *If* $G(f^-) = A_0[b_1t^{-w(1)},\ldots,b_st^{-w(s)},u]$, *we term* $b_1,\ldots,b_s$, *a set of generators of* f^-,b_j *having weight* $w(j)$. *If, further,* $w(j) = -f(b_j)$, *we refer to the set* $b_1,\ldots,b_s$ *as a standard set of generators.*

Then we have the following companion to Theorem 2.35, stated without proof.

THEOREM 2.36. *Let* $b_1,\ldots,b_s$ *be a set of generators of* f^- *with weights* $w(1),\ldots,w(s)$. *Then the* A_0*-module* $I_{-n}(f)$ *consisting of all elements* x *of* A *such that* $f(x) \geq -n$ *is generated by the set of all products of elements* b_j *of weight* $\leq n$.

4. Miscellaneous results.

In this section we collect together a number of results which we will use later.

First we consider extensions of noether filtrations. Let f be a noether filtration on a ring A and let A_0 be the sub-ring of A consisting of all elements x of A such that $f(x) \geq 0$. Then, by Theorems 2.35, 2.36, and the arguments preceding them, A is a finitely generated extension $A_0[b_1,...,b_s]$ of A_0, and

$$G(f) = A_0[t^{-w'(1)}b_1,...,t^{-w'(s)}b_s,u\ ;\ t^{w(1)}a_1,...,t^{w(r)}a_r],$$

where $w(i) = f(a_i)$ and $w'(j) = -f(b_j)$. Now, suppose we have a homomorphism $\phi:A \rightarrow B$ and a noether sub-ring B_0 of B such that $B_0 \supseteq \phi(A_0)$ and $B = B_0[\phi(b_1),...,\phi(b_s)]$. Then the graded ring

$$G(fB) = B_0[t^{-w'(1)}\phi(b_1),...,t^{-w'(s)}\phi(b_s),u;t^{w(1)}\phi(a_1),...,t^{w(r)}\phi(a_r)]$$

determines a noether filtration on B which we will term the extension of f to B and denote by fB. Note, however, that the restriction of fB to A (that is, the composition of the map ϕ and the map fB of B into the real numbers plus ∞) is not, in general, the same as f, and need not be a noether filtration. A simple example is as follows. Take f to be a noether filtration on a noether domain A which takes only values ≥ 0, and satisfies the condition $f(1) = 0$, and ϕ to be the canonical map of A onto $B = A/\mathbf{p}$, where $\mathbf{p}$ is a prime ideal of A such that the union of the radical of f and $\mathbf{p}$ does not contain 1. Then, clearly fB is what we have previously denoted by $f|\mathbf{p}$. Its restriction f' to A then satisfies $f'(1) = 0$, and takes the value k on $\mathbf{p}$. It follows that it is not equal to f and cannot be noetherian since, by Theorem 2.33i), a noether filtration f with $f(1) = 0$ on a noether domain satisfies $f(x) < \infty$ for all non-zero x. Perhaps the most important new case we have to consider is that when B_0 is a ring of fractions $(A_0)_S$ of A_0, where S is a multiplicatively closed sub-set of A_0. In this case we will write f_S for fA_S.

We collect together some elementary results concerning the filtrations f_S in the following lemma.

LEMMA 2.41. *Let* f *be a noether filtration on* A, S *be a multiplicatively closed sub-set of* A_0 *and let* m *be an integer. Then, if* $x = y/s'$ *belongs to* A_S, $f_S(x) \geq m$ resp.

$(f_S)^*(x) \geq m$, *if and only if there exists* s *in* S *such that* $f(sy) \geq m$, resp. $f^*(sy) \geq m$.

We have $f_S(y/s') \geq m$ if and only if $y/s' \in I_m(f)A_S$. But this is the case if and only if, for some s, sy belongs to $I_m(f)$.

$(f_S)^*(y/s') \geq m$ if and only if $x = y/s'$ satisfies an equation

$$x^n + c_1x^{n-1}+\ldots+ c_n = 0$$

with c_j in $I_{mj}(f)A_S$. This implies that

$$s^n(y^n+s'c_1y^{n-1}+\ldots+(s')^nc_n) = 0$$

for some s in S, and we can further suppose that s is chosen so that sc_j is in $I_{mj}(f)$ for each j. Hence $f^*(sy) \geq m$. The converse is immediate.

For our main application we first require a series of lemmas.

LEMMA 2.42. *Let* A *be a noetherian domain,* $\mathbf{p}$ *be a prime ideal of* A, *and* x,y *be elements of* $\mathbf{p}$ *which are not zero divisors. Then* $\mathbf{p}$ *is an associated prime ideal of* yA *if and only if it is an associated prime ideal of* xA.

Since $\mathbf{p}$ is associated with xA, (yA) if and only if $\mathbf{p}A_{\mathbf{p}}$ is associated with $xA_{\mathbf{p}}$, $(yA_{\mathbf{p}})$, we can assume that A is local with $\mathbf{p}$ as maximal ideal. Then if $\mathbf{p}$ is associated with xA, there exists z not in xA such that $xA \supseteq z\mathbf{p}$. This implies that $\mathbf{p}^{-1} \neq A$, and hence that there exists w not in yA such that $yA \supseteq w\mathbf{p}$. Hence $\mathbf{p}$ is associated with yA. The converse follows by symmetry.

LEMMA 2.43. *Let* f *be a filtration on* A *which only takes values* ≥ 0. *Let* $\mathfrak{P}_1,\ldots,\mathfrak{P}_s$ *be the prime ideals associated with* uG(f) *and let* $\mathbf{p}_i = \mathfrak{P}_i \cap A$. *Then the set of prime ideals associated with* $I_n(f)$ *for some* n *is* $\mathbf{p}_1,\ldots,\mathbf{p}_s$ *and so is finite.*

Consider the ring G(f). Let $\mathfrak{P}_1,\ldots,\mathfrak{P}_s$ be as above. Then they are also the prime ideals associated with $u^nG(f)$ for any n. Hence we have an irredundant primary decomposition

$$u^nG(f) = \mathbf{Q}_1\cap\ldots\cap\mathbf{Q}_s$$

with $\mathbf{Q}_i$ a $\mathfrak{P}_i$-primary ideal. Then if $\mathbf{p}_i = \mathfrak{P}_i\cap A$,and $\mathbf{q}_i = \mathbf{Q}_i\cap A$, we have a primary decomposition

$$I_n(f) = \mathbf{q}_1 \cap \ldots \cap \mathbf{q}_s$$

of $I_n(f)$ with $\mathbf{q}_i$ $\mathbf{p}_i$-primary. Hence the associated prime ideals of $I_n(f)$ are contained in the set $\mathbf{p}_1,\ldots,\mathbf{p}_s$ for all n. In fact, $\mathbf{p}_i$ is associated with $I_n(f)$ for some n. For we can, as before, assume that A is local and $\mathbf{p}_i$ is its maximal ideal. We may then take $\mathbf{p}_i = \mathcal{P}_i \cap A$ with $\mathcal{P}_i$ maximal in the set of prime ideals associated with uG(f). Since $\mathcal{P}_i$ is associated with uG(f), there exists an element xt^n of G(f) not in uG(f) such that $uG(f) \supseteq xt^n.\mathcal{P}_i$. But x does not belong to $I_{n+1}(f)$ whereas $I_{n+1}(f) \supseteq x\mathbf{p}_i$. Hence $\mathbf{p}_i$ is associated with $I_{n+1}(f)$.

DEFINITION. *The finite set of prime ideals associated with* $I_n(f)$ *for some* n *is termed the set of prime ideals associated with* f.

We now use the above to prove a theorem which allows us, essentially, to reduce problems concerning noether filtrations on a noetherian ring A to the case where A is local. First we consider some matters of notation. Suppose that $\mathbf{p}$ is a prime ideal of A and f is a noether filtration on A. Then if S is the set of elements of A not in $\mathbf{p}$, we will denote f_S by $f_{\mathbf{p}}$. Note that $f_{\mathbf{p}}$ is a noether filtration on the local ring $A_{\mathbf{p}}$. Further, if x belongs to A, then $f_{\mathbf{p}}(x)$ will be the value of $f_{\mathbf{p}}$ at the image of x in $A_{\mathbf{p}}$.

THEOREM 2.44. *Let* A *be a noetherian ring,* f *a noether filtration on* A *taking values* ≥ 0. *Let* $\mathbf{p}_1,\ldots,\mathbf{p}_k$ *be prime ideals of* A *containing the radical of* f *such that* $\mathbf{p}_i$ *does not contain* $\mathbf{p}_j$ *if* $i \neq j$. *Assume also that every prime ideal* $\mathbf{p}$ *associated with* $I_n(f)$ *for some* n *is contained in at least one of the prime ideals* $\mathbf{p}_i$. *Then, writing* $f_i(x)$ *for* $f_{\mathbf{p}}(x)$ *if* $\mathbf{p} = \mathbf{p}_i$,

$$f(x) = \mathrm{Min}\, f_i(x) \quad \textit{for all } x \textit{ in } A$$

the minimum being taken over $i = 1,\ldots,k$.

Let S be the set of elements not in any of the prime ideals $\mathbf{p}_i$. Then every element s of S is prime to $u^nG(f)$ for all n, and hence $f_u(x) = f_u(sx)$ for all x in G(f). Hence it follows that $f_S(x) = f(x)$ for all x in A. Now A_S is a semi-local ring with maximal ideals $\mathfrak{m}_i = \mathfrak{p}_iA_S$. Replacing A by A_S, f by f_S and $\mathfrak{p}_i$ by $\mathfrak{m}_i$, we can assume that A is semi-local and that $\mathfrak{p}_1,\dots,\mathfrak{p}_k$ are its maximal ideals. But then, if A_i denotes the localisation of A at $\mathfrak{p}_i$, f_i is the restriction of fA_i to A. Hence, if $f_i(x) \geq m$ for all i, there exists s_i not in $\mathfrak{p}_i$ for each i such that $s_ix \in I_m(f)$, i.e., $I_m(f){:}x$ is not contained in $\mathfrak{p}_i$. Since this is true for each i, it follows that $I_m(f){:}\ x = A$ and hence $f(x) \geq m$. But clearly $f_i(x) \geq f(x)$ for each i and hence the result follows.

We conclude this section by considering the filtrations equivalent to a given noether filtration f. First we remark that, in general, such a filtration need not be noether. We use the following simple construction. Let f be a noether filtration taking only values ≥ 0. We now construct a new filtration g on A by defining g(x) to be 0 if f(x) = 0 and to be f(x) − 1 if f(x) > 0. Clearly g is a filtration equivalent to f, and has the property that,if g(x),g(y) are finite and positive,then $g(xy) > g(x) + g(y)$.

LEMMA 2.45. *If a noether filtration* g(x) *on* A *taking only non-negative values has the property that* $g(xy) > g(x) + g(y)$ *whenever* g(x),g(y) *are positive and finite, then* g(x) *takes only a finite set of positive finite values.*

Suppose that $a_1,\dots,a_{n+1}$ are elements of A such that $1 \leq f(a_1) < f(a_2) < \dots < f(a_{n+1}) < \infty$. Write $f(a_r) = m(r)$. Then the elements $a_rt^{m(r)}$ belong to $G^+(g)$. But $a_{n+1}t^{m(n+1)}$ does not belong to the ideal generated by $a_rt^{m(r)}$ for r = 1,...,n, since this would imply that $a_{n+1}t^{m(n+1)} = \Sigma a_rt^{m(r)}b_rt^{m(n+1)-m(r)}$ with $g(b_r) \geq m(n+1) - m(r)$. Then

$$a_{n+1} = \Sigma a_rb_r$$

and the condition on g now implies that $g(a_{n+1}) > m(n+1)$ contrary to hypothesis. Now if g(x) took an infinite set of positive values, we could construct an infinite sequence of elements a_r such that $1 \leq g(a_r) = m(r) < g(a_{r+1}) = m(r+1)$ for all r, and the ideal of $G^+(g)$ generated by the elements $a_rt^{m(r)}$ would not be finitely generated. Hence $G^+(g)$ is not noetherian and the same is true of g.

It follows that a noetherian filtration f is equivalent to a non-noetherian filtration g if f(x) takes an infinite set of finite positive values, a mild restriction on f (satisfied, for example, by the filtration f_{aA} on A if a is neither a unit nor a zero divisor).

In the other direction we have the following equally simple result.

LEMMA 2.46. *Let* f(x) *be a noether filtration on a ring* A, *and let* g(x) *be a filtration equivalent to* f(x) *and satisfying* $g(x) \geq f(x)$ *for all* x. *Then* g(x) *is a noether filtration.*

$A[t,u] \supseteq G(g) \supseteq G(f)$, and, for some integer k, $u^{-k}G(f) \supseteq G(g)$. Hence, as G(f) is noetherian, G(g) is a finite G(f)-module and so is noetherian. Hence g is a noether filtration.

3. THE THEOREMS OF MATIJEVIC AND MORI-NAGATA

1. Matijevic's Theorem.

This section is devoted to the proof of one of the most beautiful results of recent years in commutative algebra and some of its consequences. This theorem is due to J.R. Matijevic (Matijevic[1976]). The treatment given below is based on Kiyek[1981].

LEMMA 3.11. *Let* A *be a noetherian ring, and let* M, N *be two* A*-modules such that* M *is a sub-module of* N *and*

i) M *is finitely generated,*

ii) N *is an essential extension of* M, *i.e., every non-zero sub-module of* N *has non-zero intersection with* M,

iii) *for all* x *in* N, $l((Ax+M)/M) < \infty$, *where* l *is the length function.*

Then, if a *is a non-zero divisor of* M, (*and hence* N), N/aN *is finitely generated.*

We first observe that, if N' is a finitely generated sub-module of N, then $l((M+N')/M) < \infty$.

Next consider the descending sequence of modules $(a^mAx+M)/M$, where x is an element of N. These all have finite length, and hence, for some m depending on x,

$$a^mAx + M = a^{m+1}Ax + M = \dots$$

Our first aim is to show that m can be chosen independent of x. Now define M_r to be the sub-module $aM + (a^rN\cap M)$ of M. The sequence of modules M_r/aM is descending. The first, M_1/aM, is of finite length since it is finitely generated by, say, the images of $ax_1,\dots,ax_s$, with $x_1,\dots,x_s$ in N, and, if N' is the sub-module of N generated by $x_1,\dots,x_s$, then (M+N')/M has finite length. Hence the same is true of $(aN'+aM)/aM = M_1/aM$. It now follows that there exists an integer n such that

$$M_n = M_{n+1} = \dots$$

Now we show that, for all x in N, $a^nx \in a^{n+1}N + M$. For if this is not true, the least integer s such that $a^sx \in a^{s+1}N + M$ is greater than n for some x. Then

$$a^sx = a^{s+1}y + z,\ \ z \in M, y \in N; \qquad a^s(x - ay) = z;$$

and z belongs to $M\cap a^sN$ which, in turn, is contained in $(M\cap a^{s+1}N) + aM$, since $s > n$.

Then

$$a^s(x - ay) = a^{s+1}y' + az', \quad z' \in M, y' \in N;$$
$$a^{s-1}x = a^s(y + y') + z'$$

contrary to the definition of s. Therefore

$$a^{n+1}N + M \supseteq a^nN$$

and hence $a^nN/a^{n+1}N$ is isomorphic to a sub-module of $(a^{n+1}N + M)/a^{n+1}N$. But this is isomorphic to the finitely generated module $M/M \cap a^{n+1}N$. Hence N/aN, which is isomorphic to $a^nN/a^{n+1}N$, as a is not a zero divisor on N, is finitely generated.

Now we come to Matijevic's Theorem, which we state below for A an integral domain and in a slightly amplified form. First however we require a simple lemma.

LEMMA 3.12. *Let* A,B ⊇ A, *be noetherian domains,* B *being a finitely generated* A-*module. Then* B *is a field if and only if* A *is a field.*

Since B is a finitely generated A-module, any non-zero element x of B is integrally dependent on A and so satisfies an equation

$$x^n + a_1x^{n-1} + \ldots + a_n = 0, \quad a_i \in A,$$

which we assume to be of lowest possible degree. Then as B is a domain, $a_n \neq 0$. Hence if A is a field, so that a_n has an inverse, x has an inverse in B. Suppose that B is a field and that u is a non-zero element of A. Consider the sub-A-module M of B generated by $u^{-1}, u^{-2}, \ldots$ This is finitely generated and hence there exists n such that

$$u^{-n-1} = a_1u^{-n} + \ldots + a_nu^{-1}$$

with all a_i in A. Then, multiplying by u^n, we see that u^{-1} is in A, i.e., A is also a field.

THEOREM 3.13. *Let* A *be a noetherian domain and let* T(A) *be the set of elements* x *of the field of fractions* F *of* A *which satisfy the condition that* xI ∈ A *for some ideal* I *of* A *such that* A/I *has finite length. Let* B *be any ring such that* T(A) ⊇ B ⊇ A. *Then the following statements are true:*

i) B *is a noetherian ring.*

ii) *A non-zero prime ideal* $\mathbf{p}'$ *of* B *is maximal if and only if* $\mathbf{p} = \mathbf{p}' \cap A$ *is maximal.*

iii) *If* $\mathbf{p}$ *is a maximal ideal of* A, *there exist only a finite set of prime ideals* $\mathbf{p}'$ *of* B *such that* $\mathbf{p}' \cap A = \mathbf{p}$, *and, for each such prime ideal,* $B/\mathbf{p}'$ *is a finite algebraic extension of* $A/\mathbf{p}$.

iv) *If* $\mathbf{p}$ *is a non-maximal prime ideal of* A, *then* $A_{\mathbf{p}} \supseteq B$. *Further, if* B *is not a field, the correspondence between the non-maximal prime ideals* $\mathbf{p}'$ *of* B *and the non-maximal prime ideals* $\mathbf{p}$ *of* A *given by*

$$\mathbf{p} = \mathbf{p}' \cap A$$

is 1-1 and has inverse

$$\mathbf{p}' = \mathbf{p}A_{\mathbf{p}} \cap B$$

and further, $A_{\mathbf{p}} = B_{\mathbf{p}'}$.

i) Let J be a non-zero ideal of B and let a be a non-zero element of $A \cap J$. Then, by Lemma 3.11, J/aJ is a finitely generated A-module, whence J is a finitely generated ideal of B.

ii) If $\mathbf{p}'$ is a non-zero prime ideal of B, then $\mathbf{p} = \mathbf{p}' \cap A$ is a non-zero prime ideal of A and, by Lemma 3.11, $B/\mathbf{p}'$, being a homomorphic image of B/aB for a non-zero a in $\mathbf{p}$, is a finitely generated A-module, and therefore a finitely generated $A/\mathbf{p}$-module. It follows from the lemma above that $B/\mathbf{p}'$ is a field if and only if $A/\mathbf{p}$ is a field, i.e., $\mathbf{p}'$ is maximal if and only if $\mathbf{p}$ is maximal. (Note that the assumption that $\mathbf{p}' \neq (0)$ is essential, as the following example shows. Let A be a discrete valuation ring. Then T(A) = F, and if we take B = T(A), then (0) is a maximal prime ideal of B, but (0) is not a maximal ideal of A. I am indebted to R.Y. Sharp for pointing this out to me.)

iii) If $\mathbf{p}$ is a maximal ideal of A, then $B/\mathbf{p}B$ is a finitely generated $A/\mathbf{p}$-module and hence a finite-dimensional vector space over the field $A/\mathbf{p}$. Hence it is an artinian ring and so has only a finite number of prime ideals and the residue fields are finite extensions of the field $A/\mathbf{p}$. Both statements of iii) now follow.

iv) If $\mathbf{p}$ is a non-maximal prime ideal of A, then the definition of T(A) implies that, if x belongs to T(A), then there exists s in A not in $\mathbf{p}$ such that $sx \in \mathbf{A}$, that is, x belongs to $A_{\mathbf{p}}$. Hence $A_{\mathbf{p}} \supseteq T(A) \supseteq B$. The assumption that B is not a field implies that there exist non-maximal prime ideals of B. Suppose that $\mathbf{p}'$ is a non-maximal prime ideal of B. Then $\mathbf{p} = \mathbf{p}' \cap A$ is a non-maximal prime ideal of A by ii). Next, if x belongs to $\mathbf{p}'$, sx belongs to $A \cap \mathbf{p}' = \mathbf{p}$ and hence sx belongs to $\mathbf{p}A_{\mathbf{p}}$. Hence x belongs to $\mathbf{p}A_{\mathbf{p}}$. Hence $\mathbf{p}'$ is contained in $\mathbf{p}A_{\mathbf{p}} \cap B$. Conversely, suppose that x belongs to $\mathbf{p}A_{\mathbf{p}} \cap B$. Then there exists s in A not in $\mathbf{p}$ and therefore not in $\mathbf{p}'$ such that sx belongs to $\mathbf{p}$ and hence to $\mathbf{p}'$. It follows that x belongs to $\mathbf{p}'$. Hence $\mathbf{p}' = \mathbf{p}A_{\mathbf{p}} \cap B$ and this proves that the correspondence given between non-maximal prime ideals is 1-1 and has the given inverse. Finally the statement that $A_{\mathbf{p}} = B_{\mathbf{p}'}$ follows since $A_{\mathbf{p}}$ contains B and $\mathbf{p}' = \mathbf{p}A_{\mathbf{p}} \cap B$.

COROLLARY. $T(T(A)) = T(A)$.

Let x belong to $T(T(A))$ and let I' be the ideal of T(A) consisting of all z such that $zx \in T(A)$. Then $T(A)/I'$ is a T(A)-module of finite length and, since T(A) is noetherian, I' contains a finite product of maximal ideals of T(A). Hence $I = I' \cap A$ contains a finite product of maximal ideals of A by 3.13ii). Now xI is a finite A-module contained in T(A), and hence there exists an ideal J of A such that $A \supseteq xIJ$, and A/J has finite length. But then IJ contains a finite product of maximal ideals of A, i.e., A/IJ has finite length,and hence $x \in T(A)$. Since $T(T(A)) \supseteq T(A)$,we have proved that $T(T(A)) = T(A)$.

The following lemma is taken from Querre[1979]. Note that symbols $I : J$, where I,J are ideals of A or B, are to be interpreted as fractional ideals, i.e., they consist of all elements x of the field of fractions F of A or B such that $xJ \in I$.

DEFINITION. *A proper ideal* J *of* A *is said to be of grade* 1 *if* $J^{-1} = A : J$ *properly contains* A *and to be of grade* >1 *if* $J^{-1} = A$.

We can rephrase this definition in the form that, if a is any non-zero divisor of J, then J has grade 1 if $aA : J \neq aA$. Here we can interpret $aA : J$ as meaning the ideal of A consisting of all x in A such that $aA \supseteq xJ$, and A need not be assumed to be a domain. Note that if A is noetherian and $\mathbf{p}$ is a height 1 prime ideal containing a non-zero divisor a, then $\mathbf{p}$ is minimal over aA, and hence $aA : \mathbf{p} \neq aA$, i.e., $\mathbf{p}$ has grade 1. The full definition of grade will be given in chapter 6, but the above will be sufficient for our present purpose.

LEMMA 3.14. *With the notation of the last theorem, take* $B = A^* \cap T(A)$, *where* A^* *is the integral closure of* A *in* F. *Let* $\mathbf{p}'$ *be a maximal ideal of* B. *Then either*

a) $aB : \mathbf{p}' = aB$ *for any* $a \in \mathbf{p}'$ *(and* $\mathbf{p}'$ *has grade and height* >1*), or*

b) $B_{\mathbf{p}'}$ *is a discrete valuation ring (and* $\mathbf{p}'$ *has grade and height* 1*).*

The condition $aB : \mathfrak{p}' = aB$ implies that $\mathfrak{p}'$ is not minimal over aB and hence has height >1. Further, if it is true for a, then $\mathfrak{p}'^{-1} = B$ and hence $xB : \mathfrak{p}' = xB$ for all x in $\mathfrak{p}'$.

Now suppose that a) does not hold. Consider $\mathfrak{p}' : \mathfrak{p}'$. This is a ring containing B and integral over it and is therefore contained in A* (since B is integral over A). On the other hand, if $\mathfrak{p} = \mathfrak{p}' \cap A$,

$$T(A) \supseteq B : \mathfrak{p} \supseteq \mathfrak{p}' : \mathfrak{p} \supseteq \mathfrak{p}' : \mathfrak{p}'$$

and $\mathfrak{p}' : \mathfrak{p}' = B$. Now $B : \mathfrak{p}' \neq B$ by hypothesis. Hence $(B : \mathfrak{p}')\mathfrak{p}'$ is not contained in $\mathfrak{p}'$. This implies that $(B : \mathfrak{p}')\mathfrak{p}' = B$, since $\mathfrak{p}'$ is maximal, and hence $\mathfrak{p}'$ is invertible. This implies that $\mathfrak{p}'B_{\mathfrak{p}'}$ is a principal ideal and hence that $B_{\mathfrak{p}'}$ is a discrete valuation ring. Hence $\mathfrak{p}'$ has height 1.

The theorem following is a corollary of Querre's Lemma and generalises the Krull-Akizuki Theorem.

THEOREM 3.15. *If* A *is local with maximal ideal* $\mathfrak{m}$ *and* B *is the integral closure of* A *in* T(A), *then*

i) B *is a semi-local noetherian domain with maximal ideals* $\mathfrak{n}_1, \ldots, \mathfrak{n}_k, \ldots, \mathfrak{n}_{k+l}$, *so numbered that* $\mathfrak{n}_i$ *has height* 1 *if* $i \le k$ *and height* >1 *if* $i > k$. *Further, the localisation of* A *at* $\mathfrak{n}_i$ *is a discrete valuation ring* O_i *with associated valuation* v_i *if* $i \le k$, *while the joint localisation* C *of* B *at* $\mathfrak{n}_{k+1}, \ldots, \mathfrak{n}_{k+l}$ *contains* T(A).

ii) $B = T(A) \cap O_1 \cap \ldots \cap O_k$.

i) That B is noetherian and semi-local is an immediate consequence of Matijevic's Theorem, while the statement about the localisations at $\mathfrak{n}_1, \ldots, \mathfrak{n}_k$ is an immediate consequence of Querre's Lemma. Now suppose that $\mathfrak{n} = \mathfrak{n}_i$ with $i > k$. Then, by the above, $B : \mathfrak{n} = B$. Thus $B_{\mathfrak{n}} : \mathfrak{n}B_{\mathfrak{n}} = B_{\mathfrak{n}}$. If x belongs to T(A), then, for some n, $A \supseteq x\mathfrak{m}^n$ and hence $B_{\mathfrak{n}} \supseteq x\mathfrak{m}^n B_{\mathfrak{n}}$. As $\mathfrak{m}B_{\mathfrak{n}}$ is $\mathfrak{n}B_{\mathfrak{n}}$-primary, it follows that $x \in B_{\mathfrak{n}} : \mathfrak{n}^r B_{\mathfrak{n}}$ for some r and hence $\in B_{\mathfrak{n}}$. Since C is the intersection of the local rings $B_{\mathfrak{n}}$, where $\mathfrak{n} = \mathfrak{n}_i$, $i = k+1, \ldots, k+l$, C contains T(A).

ii) Since $T(A) \supseteq B$, it follows that B is the intersection of T(A) and the local rings $B_{\mathfrak{p}}$, where $\mathfrak{p}$ ranges over the height 1 maximal prime ideals of B, that is,

$$B = T(A) \cap O_1 \cap \ldots \cap O_k.$$

COROLLARY (The Krull-Akizuki Theorem). *If* A *is a noetherian local domain of dimension* 1 *with maximal ideal* $\mathfrak{m}$, *residue field* k *and field of fractions* F *and* A^* *is the integral closure of* A *in a finite algebraic extension* E *of* F, *then* A^* *is a semi-local noetherian domain whose maximal ideals* $\mathfrak{n}_1,\ldots,\mathfrak{n}_k$ *all have height* 1. *Further, the localisation of* A^* *at* $\mathfrak{n}_i$ *is a discrete valuation ring* O_i, *whose residue field* K_i *is a finite algebraic extension of* k. *Finally, if* v *is an integer-valued valuation on* $F \geq 0$ *on* A *and* >0 *on* $\mathfrak{m}$, *and* w *is an extension of* v *to* E *then* O_w *is one of the rings* O_i.

Let $u_1,\ldots,u_n$ be a basis of E over F. Then there exists $a \neq 0$ in A such that au_i is integral over A for each i. Then $B = A[au_1,\ldots,au_n]$ is a finite A-module, hence integral over A, and has E as field of fractions. Further, if $x \neq 0 \in A$, B/xB is a finite A/xA module and, since the latter has finite length, has finite length as an A-module. But, as B is integral over A, every non-zero prime ideal of B meets A in $\mathfrak{m}$ and so contains x. Hence B has only a finite set of maximal ideals, $\mathfrak{p}_1,\ldots,\mathfrak{p}_s$, these are maximal and minimal over xB, and $B/\mathfrak{p}_j$ is a finite extension of k for each j. It follows that all non-zero prime ideals of B have height 1. Since B has dimension 1, all non-zero ideals of B have finite co-length, and hence $T(B) = F$. Finally $A^* = B^*$.

Except for the last two statements, the whole of this corollary follows from Theorem 3.15 applied to the localisations of B at the prime ideals $\mathfrak{p}_j$. The statement that K_i is a finite algebraic extension of k follows from Theorem 3.13 iii), and the fact that, if the maximal ideal of O_i meets B in $\mathfrak{p}_j$, then $B/\mathfrak{p}_j$ is a finite extension of k. The last sentence is proved as follows. Clearly $w(x) \geq 0$ on A^*, and its centre on A^* is one of the ideals $\mathfrak{n}_j$. Suppose it is $\mathfrak{n}_i$. Then $O_v \supseteq O_i$ and, since both are discrete valuation rings and therefore maximal proper sub-rings of F, they are equal.

2. The Mori-Nagata Theorem.

Our starting point in this section is Querre's Lemma (Lemma 3.14) and its corollary, Theorem 3.15. Let A be a noetherian domain and let $\mathfrak{p}$ be a prime ideal of A.

Then this corollary associates with the local ring $A_{\mathbf{p}}$ a finite set of discrete valuations $v_1,\dots,v_k$ on the field of fractions F of A, namely those determined by the discrete valuation rings $O_1,\dots,O_k$ of the corollary, applied to the local ring $A_{\mathbf{p}}$. Clearly, $v_i(x) \geq 0$ on A and >0 on $\mathbf{p}$.

DEFINITION. *The valuations* $v_1,\dots,v_k$ *defined above are termed the Krull valuations of* A *centre* $\mathbf{p}$. *A discrete integer-valued valuation* v *on* F *which is* ≥ 0 *on* A *is termed a Krull valuation of* A *if it is a Krull valuation of* A *centre* $\mathbf{p}$, *as just defined, where* $\mathbf{p}$ *is the prime ideal of* A *consisting of all* $x \in A$ *such that* $v(x) > 0$.

(Note that the last statement in the Krull-Akizuki Theorem (Theorem 3.15, Corollary) implies that any discrete valuation centre a height 1 prime ideal of A is a Krull valuation of A. In general however, there are others.)

The following theorem, which will take the rest of this section to prove, is the version of the Mori-Nagata theorem which will be proved in this chapter. Note that iv) below gives the more usual form of the definition of Krull valuations.

THEOREM 3.21. *Let* A *be a noetherian domain with field of fractions* F. *Then the set of Krull valuations of* A *has the following properties:*

i) *the residue field* K_v *of a Krull valuation* v *of* A *is a finite algebraic extension of the field* $A_{\mathbf{p}}/\mathbf{p}A_{\mathbf{p}}$, *where* $\mathbf{p}$ *is the centre of* v;

ii) *if* $x \neq 0$ *is an element of* A, *then* $v(x) \neq 0$ *for only a finite set of Krull valuations of* A;

iii) *if* x *is an element of* F, *then* x *is integrally dependent on* A *if and only if* $v(x) \geq 0$ *for all Krull valuations* v *of* A;

iv) *there is a 1-1 correspondence* $v \longleftrightarrow \mathbf{p}_v$ *between the Krull valuations* v *of* A *and the height 1 prime ideals of* A* *defined by the statements that* $\mathbf{p}_v$ *is the centre of* v *on* A* *and the valuation ring of* v *is the localisation of* A* *at* $\mathbf{p}_v$.

We will now prove in the rest of this section that the set of Krull valuations does have the four properties stated.

Proof of 3.21 i). This follows almost immediately from the last section. For the residue field K_v of v is the residue field of a local ring $B_{\mathbf{p}'}$, where $\mathbf{p}'$ is a maximal ideal of the ring $B = A_{\mathbf{p}}^{*} \cap T(A_{\mathbf{p}})$, with $\mathbf{p}$ the centre of v. Then, by Theorem 3.13 iii), it follows that K_v is a finite algebraic extension of $A_{\mathbf{p}}/\mathbf{p}A_{\mathbf{p}}$ as required.

Proof of 3.21 ii). Since the number of Krull valuations with a given centre is finite, we have to prove that the set of prime ideals $\mathbf{p}$ containing x, for which the set of Krull valuations centre $\mathbf{p}$ is non-empty, is finite. This is a consequence of the following lemma.

LEMMA 3.211. *Let* A *be a noetherian domain,* x *be a non-zero element of* A *and let* $\mathbf{p}$ *be the centre of a Krull valuation of* A *containing* x. *Then* $\mathbf{p}$ *is one of the prime ideals associated with* xA, *whence there are only a finite number of such prime ideals, and therefore only a finite number of Krull valuations* v *of* A *such that* $v(x) > 0$.

Suppose that v is a Krull valuation with $\mathbf{p}$ as centre. Then $\mathbf{p}$ is either of height 1 and so is associated with xA, or $\mathbf{p}$ has height >1 and $A_{\mathbf{p}}^{*} \cap T(A_{\mathbf{p}})$ has a maximal ideal of height 1; hence $T(A_{\mathbf{p}}) \neq A_{\mathbf{p}}$. But this implies that $A_{\mathbf{p}} : \mathbf{p}A_{\mathbf{p}} \neq A_{\mathbf{p}}$ and hence $\mathbf{p}A_{\mathbf{p}}$ is associated with $xA_{\mathbf{p}}$ and therefore $\mathbf{p}$ is associated with xA for all x in $\mathbf{p}$. Since there are only a finite number of prime ideals associated with xA, there are only a finite number of centres of Krull valuations of A which contain x and hence only a finite number of Krull valuations v of A such that $v(x) > 0$.

Proof of 3.21 iii). Since a discrete valuation ring O_v is integrally closed, any such ring containing A contains A*, i.e., if v is a discrete valuation ≥0 on A, then v ≥ 0 on A*. Therefore all we have to show is that if z is an element of F which is not integrally dependent on A then there is a Krull valuation v on A such that $v(z) < 0$. Now we may write z in the form y/x where x,y both belong to A. It follows that z belongs to the ring of fractions $A_{(x)}$ and hence $v(z) \geq 0$ except, possibly, when v is one of the Krull valuations of A for which $v(x) > 0$. Hence 3.21 iii) will follow from the following theorem which will also be used in the next chapter.

THEOREM 3.22. *Let* A *be a noetherian domain,* x *be a non-zero element of* A *and let* $v_1,\dots,v_t$ *be the Krull valuations of* A *such that* $v_i(x) > 0$. *Then an element* z *of* $A_{(x)}$ *belongs to* $C = A^{*} \cap A_{(x)}$ *if and only if* $v_i(z) \geq 0$, $i = 1,\dots,t$.

Before proceeding to the proof of this theorem we need a lemma.

LEMMA 3.221. *Let A be a noetherian domain, x be a non-zero element of A, let* $\mathbf{p}_1,\dots,\mathbf{p}_s$ *be the prime ideals maximal among those associated with* xA *and let* Q_i *denote the localisation of A at* $\mathbf{p}_i$. *Then, if z is an element of* $A_{(x)}$ *which is integrally dependent on each of the local rings* Q_i, *it is integrally dependent on A.*

Let $z = y/x^m$. Then, since z is integrally dependent on each of the rings Q_i, it follows that we can find k such that z^n belongs to $x^{-k}Q_i$ for all i,n. Now suppose that w is an element of $x^{-k}Q_1 \cap \dots \cap x^{-k}Q_s \cap A_{(x)}$. Let J be the ideal of A consisting of all elements a such that aw belongs to $x^{-k}A$. Then J is not contained in any of the prime ideals $\mathbf{p}_1,\dots,\mathbf{p}_s$ and hence there exists an element b of A not in any of the prime ideals $\mathbf{p}_1,\dots,\mathbf{p}_s$ such that bw belongs to $x^{-k}A$. But w is of the form $x^{-r}u$ with u in A, and we can assume that $r > k$. Hence bu belongs to $x^{r-k}A$ and, since b is prime to xA and hence to $x^{r-k}A$, it follows that u is in $x^{r-k}A$. Hence w is in $x^{-k}A$. Applying this to the elements z^n, we see that all the elements z^n belong to $x^{-k}A$ and hence z is integrally dependent on A.

Proof of Theorem 3.22. We commence with a general observation. Let $\mathbf{p}$ be a prime ideal of A. Then a Krull valuation of A centre $\mathbf{p}$ is the same thing as a Krull valuation of $A_{\mathbf{p}}$ centre $\mathbf{p}A_{\mathbf{p}}$, this being an immediate consequence of the definition and the discussion preceding it. Further, since a localisation of $A_{\mathbf{p}}$ at a prime ideal coincides with a localisation of A at a prime ideal $\mathbf{p}'$ contained in $\mathbf{p}$, the Krull valuations of $A_{\mathbf{p}}$ are those of A whose centre is contained in $\mathbf{p}$.

Now let $\mathbf{p}_1,\dots,\mathbf{p}_s$ be the prime ideals maximal among those associated with xA and let h be the maximum of the heights of these prime ideals. Let z be an element of $A_{(x)}$ which is not integrally dependent on A. Then we will prove, by induction on h, that $v_i(z) < 0$ for at least one of the valuations $v_1,\dots,v_t$. First we note that z is not integrally dependent on at least one of the local rings Q_i of the last lemma and further the set of Krull valuations of Q_i is contained in the set of Krull valuations of A by the preceding paragraph. Hence we can replace A by Q_i. We now write Q for Q_i, $\mathbf{m}$ for its maximal ideal, and note that $\dim Q \leq h$. We first deal with the case $h = 1$, i.e., dim$Q$ = 1. In this case $Q_{(x)}$ is the field of fractions of Q and the same is true of $T(Q)$. The result now follows from the Theorem 3.15, Corollary, the Krull-Akizuki Theorem.

Suppose that $h > 1$. Now let $B = Q^* \cap T(Q)$. Since B is integrally dependent on Q, z is not integrally dependent on B. Further, by Matijevic's Theorem (Theorem 3.13), B is noetherian, and by Theorem 3.15,

$$B = T(Q) \cap \ldots \cap O_1 \cap \ldots \cap O_q,$$

where O_j is the valuation ring of a Krull valuation v_j of Q with centre the maximal ideal of Q and the set $v_1, \ldots, v_q$ contains all such valuations. Hence v_j is among the valuations $v_1, \ldots, v_t$. Suppose that $v_j(z) \geq 0$ for the valuations associated with the rings O_j. Then z cannot be integrally dependent on $T(Q)$. For suppose it were. Then, for some integer k, $x^k z^n \in T(Q)$, for all n, and thus, as $x^k z^n \in O_j$ for all n and each j, this would imply that $x^k z^n$ belongs to B for all n and hence z is integrally dependent on B contrary to the above. The statement $T(T(Q)) = T(Q)$ of the corollary to Theorem 3.13 can be rephrased in the form that no maximal ideal of $T(A)$ is associated with a non-zero principal ideal $aT(Q)$ of $T(Q)$. Now if $\mathbf{p}'$ is a non-maximal prime ideal of $T(Q)$ and $\mathbf{p} = \mathbf{p}' \cap Q$, then $T(Q)_{\mathbf{p}'} = Q_{\mathbf{p}}$, and hence $\mathbf{p}'$ has height $<h$, i.e., all the associated primes of $xT(Q)$ have height $<h$ for all x in $\mathfrak{m}$. It follows that we can apply the inductive hypothesis to $T(Q)$ and deduce that $v(z) < 0$ for some Krull valuation v of $T(Q)$ satisfying $v(x) > 0$. But a Krull valuation of $T(Q)$ has centre a non-maximal prime of $T(Q)$ and so is a Krull valuation of Q, by iv) of Theorem 3.13.

Proof of 3.21 iv). Again we extract part of the proof as a preliminary lemma.

LEMMA 3.23. *Let* A, C, $v_1, \ldots, v_t$ *be as in the statement of Theorem 3.22. Let* v *be one of the valuations* $v_1, \ldots, v_t$, *and let* $\mathbf{p}$ *be its centre on* C. *Then* $O_v = C_{\mathbf{p}}$ *and* $\mathbf{p}$ *has height* 1.

Let $\mathbf{p}'$ be the centre of v on A, i.e., the set of x in A such that $v(x) > 0$, and let S be the set of elements of A not in $\mathbf{p}'$. Then

$$(A_S)_{(x)} = (A_{(x)})_S,\ (A_S)^* = (A^*)_S,\ (A_S)^* \cap (A_S)_{(x)} = C_S.$$

Hence, since S does not meet $\mathbf{p}$, $C_{\mathbf{p}}$ is equal to the localisation of C_S at the centre of v, whence it follows that it will be sufficient to prove the above with A replaced by $Q = A_S$ and C replaced by C_S. Note that the centre of v on Q is the maximal ideal $\mathfrak{m}$ of

Q and that v is a Krull valuation of Q. Now let $B = T(Q) \cap Q^*$ so that $C_S \supseteq B$. But the definition of Krull valuation centre $\mathfrak{m}$ of Q implies that if $\mathfrak{p}''$ is the centre of v on B then $B_{\mathfrak{p}''} = O_V$ and $\mathfrak{p}''$ has height 1. This implies that the localisation of C_S at the centre of v is also O_V, i.e., that $C_{\mathfrak{p}} = O_V$ and $\mathfrak{p}$ has height 1.

Conclusion of Proof of 3.21 iv). The above lemma implies immediately that the localisation of A* at the centre $\mathfrak{p}_V$ of v is O_V, implying that $\mathfrak{p}_V$ has height 1 and that the map of the set of Krull valuations of A into the set of height 1 prime ideals of A* in which each valuation v maps to its centre $\mathfrak{p}_V$ is 1-1 into. We have to prove that it is onto. Let $\mathfrak{p}$ be a height 1 prime ideal of A*. Then $\mathfrak{p}$ certainly contains a non-zero element a of A and so $\mathfrak{p}$ contains the radical of aA*. But, by the above, if $v_1,\dots,v_k$ are the Krull valuations of A such that $v_i(a) > 0$ for $i = 1,\dots,k$, then the radical of aA* is the set of elements such that $v_i(a) > 0$ for $i = 1,\dots,k$. Let $\mathfrak{p}_i$ denote the prime ideal of A* consisting of all elements x of A* such that $v_i(x) > 0$. Then $\mathfrak{p}$ contains the intersection of the ideals $\mathfrak{p}_1,\dots,\mathfrak{p}_k$ and hence their product. It follows that $\mathfrak{p}$ contains $\mathfrak{p}_i$ for some i. Since $\mathfrak{p}$ and $\mathfrak{p}_i$ both have height 1, they must be equal.

COROLLARY i). *If* B *is a noetherian domain contained in* F *and containing* A, *and* v *is a Krull valuation of* A *such that* $v(x) \geq 0$ *on* B, *then* v *is a Krull valuation of* B.

For $O_V \supseteq B^* \supseteq A^*$. Since the localisation of A* at the centre of v is O_V, it follows that localisation of B* at the centre of v is also O_V. Hence this centre has height 1, and v is a Krull valuation of B by Theorem 3.21 iv).

COROLLARY ii). *The set of Krull valuations of* A_S, *where* S *is a multiplicatively closed set of elements of* A, *is the set of Krull valuations* v *of* A *such that* $v(s) = 0$ *for all* s *in* S.

The inclusion $O_V \supseteq A_S$ holds if and only if $v(s) = 0$ for all s in S. Hence, if $v(s) = 0$ for all s in S, v is a Krull valuation of A_S. On the other hand, since $A_S^* = (A^*)_S$, it follows that if v has centre height 1 on A_S^* it has centre height 1 on A*. Hence every Krull valuation of A_S is a Krull valuation of A.

COROLLARY iii). *If* x *is a non-zero element of* A *and* $v_1,\dots,v_t$ *are the Krull valuations of* A *such that* $v_i(x) > 0$, *then there exist* $y_1,\dots,y_t$ *in* $A_{(x)} \cap A^*$ *such that* $v_i(y_i) > 0$, $v_j(y_i) = 0$, *if* $j \neq i$.

For the centres $\mathbf{p}_1,\dots,\mathbf{p}_t$ of $v_1,\dots,v_t$ on $C = A_{(x)} \cap A^*$ are all of height 1 and hence no one contains another. Hence we can choose y_i in $\mathbf{p}_i$ but not in $\mathbf{p}_j$ for $j \neq i$, proving the result.

COROLLARY iv). *With the notation of the last corollary, there exists a finite integral extension* A' *of* A *contained in* $A_{(x)}$, *such that* $v_1,\dots,v_t$ *have centres of height* 1 *on* A'.

We take $A' = A[y_1,\dots,y_t]$ with $y_1,\dots,y_t$ as in the last lemma. It now follows that, if $\mathbf{p}_i'$ is the centre of v_i on A', then $\mathbf{p}_i'$ is not contained in $\mathbf{p}_j'$ if $i \neq j$. Now suppose $\mathbf{p}_i'$ had height >1. Then $\mathbf{p}_i'$ contains a prime ideal $\mathbf{p}$ minimal over xA' which therefore has height 1 and so is the centre of a Krull valuation of A'. Hence $\mathbf{p}$ must be one of the prime ideals $\mathbf{p}_j'$, $j \neq i$, which is a contradiction. Hence all the ideals $\mathbf{p}_i'$ have height 1.

We now use this last corollary to give another characterisation of the Krull valuations of a noetherian domain A.

THEOREM 3.24. *Let* A *be a noetherian domain, field of fractions , and let* v *be an integer-valued valuation on* F *such that* $v(x) \geq 0$ *on* A. *Then a necessary and sufficient condition for* v *to be a Krull valuation of* A *is that there should exist a finite integral extension* B *of* A, *contained in* F, *such that the centre* $\mathbf{p}$ *of* v *on* B *is of height* 1. *Further, if* x *is any element of* A *such that* $v(x) > 0$, *then* B *can be chosen so that it is contained in* $A_{(x)}$.

The necessity of the condition, and the last sentence of the statement of the theorem, follows immediately from Corollary iv) above. To prove sufficiency,we first note that B* = A*. Hence, using the criterion that v is a Krull valuation of A if and only if its centre on A* is of height 1, we see that A,B have the same Krull valuations. But if v has centre on B a prime ideal of height 1, then v is a Krull valuation of B by the remarks following the definition of Krull valuations at the beginning of this section.

4. THE VALUATION THEOREM

1. The Valuation Theorem.

The Valuation Theorem referred to in the title of this section states, in its simplest form, that, if $f(x)$ is a noether filtration on a noetherian ring A, then

$$\mathfrak{f}(x) = \operatorname{Min} v_i(x)/e_i$$

where $v_1,\ldots,v_k$ is a finite set of integer-valued valuations on A and e_i is a positive rational number. The representation of $f(x)$ in this form that we will construct below will be irredundant, and, assuming that the valuation v_i takes all integer values, the rational number e_i is uniquely determined. However, e_i is not uniquely defined if the valuation v_i takes only the values 0 and ∞, the latter value being taken on a minimal prime ideal of A. In this case e_i may take any positive value and will usually be given the value 1.

We will prove this theorem by stages. Initially, we assume that A is a noetherian domain, and in the lemma following, restrict attention to the case where f is a principal filtration. As we shall see, the general theorem is proved by reduction to this case.

LEMMA 4.11. *Let A_0 be a noetherian domain,* u *be a non-zero element of A_0, let* A *be the ring $(A_0)_{(u)}$ and let f_u be the principal filtration defined on* A *by* u, A_0. *Finally, let $V_1,\ldots,V_k$ be the Krull valuations on A_0 such that $V_i(u) > 0$. Then, if $x \in$* **A**,

i) $f_u^*(x) \geq n$ *if and only if* $V_i(x) \geq nV_i(u)$, $i = 1,\ldots,k$,

ii) $\mathfrak{f}_u(x) = \operatorname{Min}(V_i(x)/V_i(u))$ *and this representation is irredundant,*

iii) $f_u^*(x) = [\mathfrak{f}_u(x)]$, *where the right-hand side denotes the integer part of* $\mathfrak{f}_u(x)$.

We recall that $f_u(x)$ is the greatest integer n such that $u^{-n}x$ belongs to A_0 (and is ∞ if $x = 0$). Then $f_u^*(x) \geq n$ if x satisfies an equation

$$x^r+a_1x^{r-1}+\ldots+a_r = 0$$

with $f_u(a_i) \geq ni$ for $i = 1,\ldots,r$. This condition is equivalent to $u^{-n}x \in A_0^*$. But as x belongs to $A = (A_0)_{(u)}$, it follows from Theorem 3.22 that this is the case if and only if $V_i(x) \geq nV_i(u)$ for $i = 1,\dots,k$, which proves i).

We now prove ii) and iii) together. First we can rewrite i) in the form

$$f_u^*(x) = [\mathrm{Min}\,(V_i(x)/V_i(u)],$$

the minimum being over $i = 1,\dots,k$. We now consider $\mathbf{f}_u(x)$. Write $g(x)$ for $\mathrm{Min}\,(V_i(x)/V_i(u))$. Then $g(x)$ is a homogeneous filtration and $g(x) \geq f_u^*(x) \geq f_u(x)$ for all x by Lemma 2.21. Hence $g(x) \geq \mathbf{f}_u(x)$ for all x. On the other hand, suppose the integer n is a multiple of all the integers $V_i(u)$. Then, since the right-hand side is an integer, $f_u^*(x^n) = \mathrm{Min}\,(nV_i(x)/V_i(u))$. It follows that $\mathbf{f}_u(x^n) = n.\mathrm{Min}\,(V_i(x)/V_i(u))$ and as $\mathbf{f}_u$ is homogeneous, we have,

$$\mathbf{f}_u(x) = \mathrm{Min}\,(V_i(x)/V_i(u)).$$

Let $\mathbf{p}_j$ denote the centre of V_j on $A_0^* \cap A$ for $j=1,\dots,k$. Note that the prime ideals $\mathbf{p}_1,\dots,\mathbf{p}_k$ are distinct of height 1, by Lemma 3.23. Then, using Corollary iii) to Theorem 3.22, if y_i belongs to $\mathbf{p}_j$ for all $j \neq i$ but not to $\mathbf{p}_i$, $V_j(y_i) > 0$ if $j \neq i$ but $V_i(y_i) = 0$. But we can write $y_i = x_i/u^n$ with x_i in A, for some integer n, and then $V_i(x_i)/V_i(u) = n < V_j(x_i)/V_j(u)$ if $j \neq i$. This shows that the given representation of $\mathbf{f}(x)$ is irredundant.

Our next step is to remove the restriction in Lemma 4.11 that A_0, A are domains, and for this purpose, we must define the Krull valuations of a noetherian ring, and not merely of a noetherian domain.

DEFINITION. *An integer-valued valuation* V *on a noetherian ring* A *is termed a Krull valuation of* A *if it is of the form*

$$V(x) = W(\theta_{\mathbf{p}}(x)),$$

where $\mathbf{p}$ *is a minimal prime ideal of* A, $\theta_{\mathbf{p}}$ *is the canonical map* $A \rightarrow A/\mathbf{p}$, *and* W *is a Krull valuation of* $A/\mathbf{p}$. $\mathbf{p}$ *is the set of elements* x *of* A *such that* $V(x) = \infty$, *and, as such is termed the limit ideal of* V. *(Note that any real-valued valuation* V *on a noetherian ring* A *is of the form* $W(\theta_{\mathfrak{p}}(x))$, *where* $\mathfrak{P}$ *is the limit ideal of* V *and* W *is a valuation on* $A/\mathfrak{P}$.)

With minor modifications, definitions and theorems concerning Krull valuations of noetherian rings are similar to those already given in the domain case. We list a few that we require below.

As in the domain case, the centre $\mathfrak{p}_V$ of a valuation V on A satisfying $V(x) \geq 0$ on A is the prime ideal of A consisting of all x such that $V(x) > 0$. The residue field K_V of V is the residue field of W. Finally properties i),ii) of Theorem 3.21 take the following form.

4.1A. *If* V *is a Krull valuation on* A *with centre* $\mathfrak{p}$, *then* K_V *is a finite algebraic extension of* $A_{\mathfrak{p}}/\mathfrak{p}A_{\mathfrak{p}}$.

4.1B. *If* x *is an element of* A *contained in no minimal prime ideal of* A *(in particular, if* x *is not a zero divisor), then*

$$V(x) = 0$$

for all save a finite set of Krull valuations of A.
(Note that this follows immediately if we note that A only has a finite number of minimal prime ideals).

THEOREM 4.12. *Let* A_0 *be a noetherian ring,* u *be a non-zero divisor of* A_0 *and let* $A = (A_0)_{(u)}$. *Then, if* $V_1,\ldots,V_k$ *are the extensions to* A *of the Krull valuations of* A_0 *such that* $V_i(u) > 0$,

i) $f_u^*(x) \geq n$ *if and only if* $V_m(x) \geq nV_m(u)$, $m = 1,\ldots,k$,

ii) $\mathfrak{f}_u(x) = \mathrm{Min}\,(V_m(x)/V_m(u) | m = 1,\ldots,k)$,

iii) $f_u^*(x) = [\mathfrak{f}_u(x)]$.

Let $\mathfrak{P}_1,\ldots,\mathfrak{P}_s$ be the minimal prime ideals of A_0. Then, by Lemma 2.32,

$$f_u^*(x) = \mathrm{Min}\,((f_u/\mathfrak{P}_i A)^*(x) | i - 1,\ldots,s).$$

Let u(i),x(i) denote the images of u,x in $A/\mathfrak{P}_iA$. Then $f_u/\mathfrak{P}_iA$ is the filtration $f_{u(i)}$ on $(A_0/\mathfrak{P}_i)_{u(i)} = A/\mathfrak{P}_iA$. Now let W_{ij}, $j = 1,\dots,k_i$, be the Krull valuations on $A_0/\mathfrak{P}_i$ satisfying $W_{ij}(u(i)) > 0$ for $i = 1,\dots,s$. Define $V_m(x) = W_{ij}(\theta_i(x))$ for x in A, where θ_i is the canonical map $A \to A/\mathfrak{P}_iA$ and $m = k_1 + \dots + k_{i-1} + j$, and write k for $k_1 + \dots + k_s$. It now follows from the above that $f_u{}^*(x) \geq n$ if and only if $f_{u(i)}{}^*(x(i)) \geq n$ for $i = 1,\dots,s$, and, hence, by Lemma 4.11, if and only if

$$V_m(x) \geq nV_m(u), \quad m = 1,\dots,k.$$

This proves i). Parts ii), iii), with the exception of the irredundancy statement in ii), follow as in Lemma 4.11. We now consider the irredundancy statement. We have to prove the existence, for each m, of an element y_m in A such that the statement

$$S(m,j): \qquad V_m(y_m)/V_m(u) < V_j(y_m)/V_j(u)$$

holds for all $j \neq m$. Fix m, and divide the set of values of $j \neq m$ into two sets S, S', S being the set of values of j such that $V_m(x), V_j(x)$ take the value ∞ on the same minimal prime ideal $\mathfrak{P}$ of A, and S' consisting of all other values of j. Then it follows from Lemma 4.11 that we can find y such that S(m,j) holds for all j in S with $y_m = y$. Next we can choose z in A not in $\mathfrak{P}$ but in all other minimal prime ideals of A, so that $V_j(z) = \infty$ if $j \in S'$, while $V_j(z)$ is finite if $j \in S$. Hence we can choose N such that

$$N\{V_m(y)/V_m(u) - V_j(y)/V_j(u)\} > \{V_j(z)/V_j(u) - V_m(z)/V_m(u)\}$$

for all j in S. It now follows that if $y_m = y^N z$, S(m,j) holds for all $j \neq m$, and the irredundancy statement follows.

Our next step is to introduce the assumption that A_0 above is a graded noetherian ring, and that u is homogeneous, implying that A is also graded. To avoid confusion, we will write G in place of A_0 and simply write $G_{(u)}$ for A.

If V is a valuation on G, then, if we write an element x in the form Σx_i, it follows that

$$V(x) \geq \operatorname{Min} V(x_i).$$

We will say that V is a *graded valuation* if equality holds for all x. We can always construct a graded valuation V_h from v by defining $V_h(x) = \text{Min } V(x_i)$. This defines a valuation, the only difficulty being the proof of the equation $V_h(xy) = V_h(x) + V_h(y)$. To see this, write $x = \Sigma x_i$, $y = \Sigma y_j$ and choose r,s to be the least integers such that $V_h(x) = V(x_r)$, $V_h(y) = V(y_s)$. Then it is easy to see that if $xy = \Sigma z_k$, then $V(z_k) \geq V_h(x) + V_h(y)$ for all k, with equality if $k = r + s$. This proves the result. Note that V_h is the "largest" graded valuation less than or equal to v(x).

LEMMA 4.13. *Let* G *be a graded noetherian ring and let* u *be a homogeneous non-zero divisor of* G, *so that* $G_{(u)}$ *is also graded. Then the valuations* V_m *of Theorem 4.12 are graded valuations.*

Since the function $f_u(x)$ is graded in the sense that $f_u(x) = \text{Min } f_u(x_i)$, it follows from $V_m(x)/V_m(u) \geq f_u(x)$ that $(V_m)_h(x)/V_m(u) \geq f_u(x)$ for all x. Hence $(V_m)_h(x)/V_m(u) \geq \boldsymbol{f}_u(x)$ for $m = 1,\dots,k$ and

$$\boldsymbol{f}_u(x) \leq \text{Min } (V_m)_h(x)/V_m(u) \leq \text{Min } V_m(x)/V_m(u) = \boldsymbol{f}_u(x).$$

It now follows from Lemma 2.12 that $V_m(x) = (V_m)_h(x)$ for each m, proving that the valuations V_m are graded.

Our next two lemmas need a fair amount of preparation, and, for simplicity, we restrict G to be a graded noetherian domain. We will also assume the existence of a non-zero homogeneous element u of G of degree −1.

Let S be the set of non-zero homogeneous elements of G and let $F = G_S$. Then $F = F_0[t,u]$, where F_0 is a field consisting of the elements of degree 0 of F, and $t = u^{-1}$ is of degree 1. Further, the set of homogeneous elements of F of degree n is F_0t^n. (Note that we could replace t by any non-zero element of F_1.) Finally any homogeneous element of F of degree $\neq 0$ is transcendental over F_0. Now let V be an integer-valued valuation on G, and denote its extension to F by the same symbol. Restrict V to satisfy the condition that V(x) takes all integer values on F (or at least one pair of consecutive integer values on G). Let v be the restriction of V to F_0 and let q = V(u). Then the fact that V is graded implies that V is uniquely determined by v and q, since, if x belongs to F_0, $V(xt^r) = v(x) - rv(u) = v(x) - rq$. (We note in passing that this will be true for any graded valuation V on G, whether integer-valued or not.) However, the valuation v need not take all integer values. First suppose that it only takes the values $0,\infty$. Since V takes all integer values on F, it follows that $q = \pm 1$ and hence, for non-zero homogeneous X, $V(X) = \pm \deg X$. If v takes as values the multiples of a positive integer m, then the restriction that V takes all integer values is equivalent to the condition that m,q are mutually prime.

It will be convenient below to assume that v either takes only the values $0,\infty$ or takes all integer values, which we will express by saying that v is normalised. In the case where v takes values a multiple of $m > 1$, this involves multiplying the values of v(x) and hence V(x) and q by m^{-1}. This implies that V is no longer integer-valued, but takes values multiples of m^{-1} and V(u) is now equal to q/m which we denote by e. By an abuse of notation, if the original integer-valued valuation V is a Krull valuation of G, we will also term the modified valuation V a Krull valuation of G.

We now summarise our first group of results on the graded case as a lemma.

LEMMA 4.14. *Let* G *be a graded noetherian domain containing a homogeneous element* $u \neq 0$ *of degree* -1. *Let* V *be a graded valuation on the field of fractions* F *of* G *whose restriction* v *to the field of fractions* F_0 *of* G_0 *either takes only the values* $0,\infty$ *or is normalised.*

i) *If* v *takes only the values* $0,\infty$, *then* $V(X) = C.\deg X$, *where* C *is a constant rational number.*

ii) *If* v *takes all integer values, then* V *takes values multiples of* m^{-1}, *where* m *is a positive integer, and* $V(u) = q/m$, *where* q *is prime to* m. *Further,* F *contains a homogeneous element* x *of non-zero degree such that* $V(x) = 0$ *and the residue field* K_V *is a finitely generated extension of* K_v *of transcendence degree* 1.

iii) *If* V *is a Krull valuation of* G, *then either*

a) *there is a homogeneous element* X *of non-zero degree such that* $V(X) = 0$,

or b) V(X) *is a multiple of* degX, *and all non-zero elements of* G *have degree* ≤ 0.

i) This is proved above.

ii) The statements of this part are proved above, except for the last sentence. Since v takes all integer values, we can choose y in F_0 such that $v(y) = -q$. Then, if $X = yu^m$, X is a homogeneous element of F of degree $-m \neq 0$, and $V(X) = 0$. Let $F' = F_0[X,X^{-1}]$. Then X is transcendental over F_0 and the restriction V' of V to F' is given by

$$V'(\Sigma c_r X^r) = \mathrm{Min}\, v(c_r).$$

Hence the valuation ring of V' is the localisation of $O_v[X,X^{-1}]$ at $\mathbf{p}O_v[X,X^{-1}]$, where $\mathbf{p}O_v$ is the maximal ideal of O_v. It follows that the residue field $K_{V'}$ of V' is a pure transcendental extension of K_v of transcendence degree 1. Now, as the mth powers of all elements of F belong to F', and G is finitely generated over G_0, F is a finite algebraic extension of F', implying that any extension of V' to F (in particular V) has residue field a finite algebraic extension of $K_{V'}$. Hence K_V is a finitely generated extension of K_v of transcendence degree 1.

iii) Let V be a Krull valuation of G. Then $V(x) \geq 0$ on G. Suppose G contains no homogeneous element x of non-zero degree such that $V(x) = 0$. Then the centre $\mathbf{p}$ of V on G contains all elements of non-zero degree. Hence, if $\mathbf{p}' = \mathbf{p} \cap G_0$, the residue field K_V of V is a finite algebraic extension of the field of fractions of $G_0/\mathbf{p}'$ and, *a fortiori*, of K_v. Then, by ii), the possibility that v takes all integer values is excluded, and so we are in case i), i.e., v(x) takes only the values $0,\infty$ on F_0 and further $V(X) = C.\deg X$, where C is rational. As V(u) is positive, C is negative, and hence G can contain no non-zero element s of positive degree.

We now take G = G(f), where f is a noether filtration on a noetherian ring A, and, as usual, let A_0 denote the sub-ring of A consisting of elements x such that $f(x) \geq 0$, so that $G_0 = A_0$. Let v be a real-valued valuation on A, taking values ≥ 0 on A_0 and the value ∞ when $f(x) = \infty$, and let a be any real number. Then, as above, there is a unique graded extension V of v to G such that $V(u) = a$. We now consider the conditions on a that $V(X) \geq 0$ on G. The condition that $a = V(u) \geq 0$ is clearly necessary. We consider separately the conditions that $V(X) \geq 0$ on $G(f^+) = G^+(f)[u]$, and on $G(f^-)$.

Consider the first of these. It is clearly sufficient to consider the condition when X is of the form xt^n with x in A_0 and $f(x) \geq n$. If n is negative, the two conditions: i) $v(x) \geq 0$ on A_0, ii) $a \geq 0$, are sufficient. If $f(x) = \infty$, then $V(X) = \infty$, whatever the value of a. Hence we may assume that $f(x) < \infty$. Now if $f(x) > n$, then $xt^{f(x)}$ belongs to $G(f)$, and the condition that $V(xt^{f(x)}) \geq 0$, together with $a \geq 0$, implies that

$$V(xt^n) = V(xt^{f(x)}.u^{f(x)-n}) = V(xt^{f(x)}) + (f(x) - n)a \geq 0.$$

Hence we may restrict attention to the condition that $V(X) \geq 0$, for X of the form xt^n where $f(x) = n < \infty$. We can write the condition in this case in the form

$$v(x)/f(x) \geq a.$$

We now introduce a definition.

DEFINITION. $v(f^+) = \text{Inf}\ \{v(x)/f(x) | 0 < f(x) < \infty\}$

Then $V(X) \geq 0$ on $G(f^+)$ if and only if $v(f^+) \geq a \geq 0$.

We can define $v(f^+)$ slightly differently. Let $a_1,\dots,a_r$ be a standard set of generators of f^+, a_i having weight $w(i) = f(a_i) < \infty$. Let $M = \text{Min}\ \{v(a_i)/f(a_i) | i = 1,\dots,r\}$, so that $M \geq v(f^+)$. Now suppose that x satisfies $0 < f(x) = n < \infty$. Then x belongs to the ideal $I_n(f)$ of A_0 generated by monomials μ in $a_1,\dots,a_r$ of weight $\geq n$, and hence $v(x) \geq v(\mu)$ for at least one such monomial. Hence $v(x) \geq nM$, i.e., $v(x)/f(x) \geq M$. Hence $v(f^+) = M$.

We now turn to the case of $G(f^-)$. First we note that, if $f(x) = -n < 0$, then $xu^r \in G(f^-)$. Hence the condition $v(x) + na \geq 0$ is a necessary and suffficent condition for $V(xu^r) \geq 0$.

We now introduce the definition.

DEFINITION. $v(f^-) = \text{Sup}\ \{-v(x)/f(x), 0 | f(x) < 0\}$.

Then, arguing in the same way as above, we obtain as a necessary and sufficient condition for $V(x) \geq 0$ on $G(f^-)$ in the form $a \geq v(f^-)$.

Again using similar arguments to the above, we can express $v(f^-)$ in terms of a standard set of generators $b_1,\dots,b_s$ of f^-, b_j having weight $w(j) = -f(b_j)$, in the form

$$v(f^-) = \text{Max}\,\{-v(b_j)/f(b_j), 0 \mid j = 1,\dots,s\}.$$

Putting these together we obtain

LEMMA 4.15. *Let* v *be a valuation on* A *taking real values on* F *and values* ≥ 0 *on* A_0. *Then,*

i) *if* $v(f^+), v(f^-)$ *are as defined above,* $v(f^-) \leq v(f^+)$;

ii) *the graded extension* V *of* v *to* A[t,u] *obtained by giving* V(u) *the value* a *is* ≥ 0 *on* G(f) *if and only if* $v(f^-) \leq a \leq v(f^+)$.

i) there is nothing to prove if either $v(f^+) = \infty$ or if $v(f^-) = 0$, the latter since, as $v(x) \geq 0$ on A_0, $v(f^+) \geq 0$. Hence we may assume that there exist elements a_i, b_j such that $v(f^+) = v(a_i)/m, v(f^-) = v(b_j)/n$, where $\infty > m = f(a_i)$, and $f(b_j) = -n < 0$. Then

$$f(a_i^n b_j^m) \geq nf(a_i) + mf(b_j) = 0,$$

whence $a_i^n b_j^m$ belongs to A_0. But, then, $v(a_i^n b_j^m) \geq 0$, which implies that $nv(a_i) + mv(b_j) \geq 0$, that is $v(f^+) \geq v(f^-)$.

The statements in ii) have already been proved.

In the following theorem, the phrase "v is a normalised valuation on A" is taken to imply the following

i) v takes values from the set of integers together with ∞,

ii) the limit ideal of v is a minimal prime ideal $\mathfrak{P} = \mathfrak{P}(v)$ of A,

iii) either $v = \partial_{\mathfrak{P}}$, where $\partial_{\mathfrak{P}}(x) = 0$ if x does not belong to $\mathfrak{P}$, and ∞ if it does, or v(x) takes at least one pair of consecutive integer values.

THEOREM 4.16. *Let* A *be a noetherian ring,* f *be a noether filtration on* A, *and let* A_0 *be the sub-ring of* A *consisting of all elements* x *such that* $f(x) \geq 0$. *Then if* $v_1,\dots,v_k$ *are the normalisations of the restrictions to* A *of the Krull valuations* $V_1,\dots,V_k$ *on* G(f) *which satisfy* $V_i(u) > 0$ *there exist positive rational numbers* e_i, $i = 1,\dots,k$, *such that*

i) $f^*(x) \geq n$ *if and only if* $v_i(x) \geq ne_i$ *for* $i = 1,\dots,k$,

ii) $\mathfrak{f}(x) = \text{Min}\, v_i(x)/e_i$, *the representation being irredundant,*

iii) $f^*(x) = [\mathfrak{f}(x)]$.

If $v_i = \partial_{\mathbf{p}}$, *then* $e_i = 1$. *Otherwise* $e_i = v_i(f^+)$ *or* $v(f^-)$. *In particular, if* $f(x) \geq 0$ *or all* x *in* A, *then* $e_i = v_i(f^+)$, *while if* $f(x) \leq 0$ *for all* x *in* A, $e_i = v(f^-)$.

Further, V_i *is the normalisation of the graded extension of* v_i *to* G(f) *defined by taking* $V_i(u) = e_i$.

We apply Theorem 4.12 with A_0, A in that theorem taken as G(f) and A[t,u]. Note that this frees A, A_0 to have the meaning given in the statement of this theorem. We now consider the filtration f_u on the ring A[t,u]. Let $v_1,\ldots,v_k$ be the normalisations of the restrictions of $V_1,\ldots,V_k$ to A. Note that the normalisation v_i is equal to the restriction of V_i if V_i only takes values $0,\infty$ on A and is obtained by multiplying the restriction by m_i^{-1} if $V_i(x)$ takes values multiples of the positive integer m_i on A.

Now suppose that V is one of the valuations $V_1,\ldots,V_k$. Then V is a graded Krull valuation of G(f) by Theorem 4.12 and Lemma 4.13. If $\mathfrak{P}$ is the limit ideal of V, then V(X) is of the form $W(\theta_{\mathfrak{P}}(X))$, where $\theta_{\mathfrak{P}}$ is the canonical map of A onto $A/\mathfrak{P}$ and W is a Krull valuation on $G(f)/\mathfrak{P}$. Now suppose that the restriction of W to $A_0/A_0 \cap \mathfrak{P}$ takes values other than $0,\infty$. Then it follows from Lemma 4.14 that there is a homogeneous element X of non-zero degree of G(f) such that $V(X) = W(\theta_{\mathfrak{P}}(X)) = 0$.

Now we note that f(x) is the restriction of $f_u(x)$ to A, and consequently $f^*(x)$ and $\mathfrak{f}(x)$ are the restrictions of f_u and $\mathfrak{f}_u$: hence, we can write conclusion i) of Theorem 4.12 in the form

i) $f^*(x) \geq n$ if and only if $v_i(x) \geq ne_i$

where $e_i = V_i(u)/m_i$ if $v_i(x)$ takes non-zero finite values and is 1 if $v_i(x)$ takes only the values $0,\infty$, that is, if $v_i = v_{\mathfrak{P}}$, where $\mathfrak{P}$ is the limit ideal of v_i. Finally we note that if $v_i(x)$ takes finite values, there is a homogeneous element xt^r of G(f) with $r \neq 0$ such that $V(xt^r) = 0$. If $r > 0$, this implies that $e_i = v_i(f^+)$, while if $r < 0$, this implies that $e_i = v_i(f^-) > 0$. Note that ii), iii), except for the irredundancy statement in ii), follow from i). We now deal with the irredundancy statement. For any integer m we have the existence of an element X, which may be assumed to be homogeneous, and so of the form xu^r, with x in A, such that $V_m(X)/V_m(u) < V_j(X)/V_j(u)$ if $j \neq m$. But this clearly implies that $v_m(x)/e_m < v_j(x)/e_j$ if $j \neq m$ which is the irredundancy statement. Finally the statements following iii) have been proved above or are immediate.

COROLLARY i). *Let f be a noether filtration on the noetherian ring A taking both positive and negative finite values, and let* $A_0 = A_0(f)$. *Then, if*

$$\mathfrak{f}(x) = \operatorname{Min} v_i(x)/e_i \text{ taken over } i = 1,\dots,k$$

is the irredundant representation of $\mathfrak{f}(x)$ *as a sub-valuation, the corresponding representation for* $\mathfrak{f}^+(x)$ *is obtained by deleting those terms* $v_i(x)/e_i$ *for which* $e_i \neq v_i(f^+)$.

$f^+(x)$ is simply the restriction of $f(x)$ to A_0. It follows that $\mathfrak{f}^+(x)$ is the restriction of $\mathfrak{f}(x)$ to A_0 and hence that the expression $\operatorname{Min} v_i(x)/e_i$ is a, possibly redundant, representation of $\mathfrak{f}^+(x)$ on A_0. Now suppose that $e_i \neq v_i(f^+)$. Since f takes finite positive values, v_i cannot be degenerate, since this would imply that $v_i(x) = 0$ for $x \neq 0$ in A_0 and hence $f(x) \leq 0$ on A_0. Hence $e_i = v_i(f^-) < v_i(f^+)$. But then

$$f(x) \leq v_i(x)/v_i(f^+) < v_i(x)/e_i$$

on A_0, and so $v_i(x)/e_i$ cannot appear in an irredundant representation of $\mathfrak{f}^+(x)$. Removing such terms, we have to prove that what is left is an irredundant representation of $\mathfrak{f}^+$. Take v_i such that $v_i(x)/e_i$ is not removed. Since $v_i(x)/e_i$ occurs in the irreducible representation of $\mathfrak{f}$, it follows that there exists y in A such that $v_i(y)/e_i < v_j(y)/e_j$ for $j \neq i$. The choice of v_i implies that there exist z in A_0 such that $v_i(z) = \mathfrak{f}(z)v_i(f^+) = e_i f(z)$ and $\mathfrak{f}(z) > 0$. Since $v_i(z)/e_i = \mathfrak{f}(z)$, $v_i(z)/e_i \leq v_j(z)/e_j$ for $j \neq i$. Now for N large $\mathfrak{f}(yz^N) > 0$ and so $w = yz^N$ belongs to A_0 and satisfies the condition that $v_i(w)/e_i < v_j(w)/e_j$ if $j \neq i$. Hence $v_i(x)/e_i$ must occur in the irredundant representation of $\mathfrak{f}^+$.

COROLLARY ii). *With the same notation as in Corollary i), the irredundant representation of* $\mathfrak{f}^-(x)$ *as a sub-valuation is obtained from that of* $\mathfrak{f}(x)$ *by deleting those terms* $v_i(x)/e_i$ *for which* $e_i \neq v_i(f^-)$ *and adding a term* $\partial(x)/1$, *if necessary.*

First we observe that $\mathfrak{f}^-(x) = \mathfrak{f}(x)$ save when $\mathfrak{f}(x) > 0$, when its value is zero. Next we observe that, if z is any element of A_0 such that $f(z) > 0$, then A is contained in $(A_0)_{(z)}$, implying that $F = F_0$. Hence we obtain a, possibly redundant, representation of $\mathfrak{f}^-$ by adding a term $\partial(x)/1$ to the irredundant representation of $\mathfrak{f}$. Now suppose that $e_i \neq v_i(f^-)$ and v_i is not degenerate. Then we must have

$$e_i = v_i(f^+) > v_i(f^-).$$

Again it follows that we can remove the term $v_i(x)/e_i$ since

$$f(x) \leq v_i(x)/\, v_i(f^-) < v_i(x)/e_i \quad \text{if} \quad f(x) \leq 0.$$

To prove that if $v_i(x)/e_i$ is not removed by the above, i.e. $e_i = v_i(f^-)$, and is not degenerate, then it occurs in the irredundant representation of $\mathfrak{f}^-(x)$, we first note that we can find $y \neq 0$ in A such that $v_i(y)/e_i < v_j(y)/e_j$ if $j \neq i$. However, since the term $\partial(x)/1$ is, by assumption, possibly added to obtain the irredundant representation of $\mathfrak{f}^-$, we need, in addition that $v_i(y)/e_i < \partial(y)/1 = 0$. Now $e_i > 0$, and $e_i = v_i(f^-)$, and hence there exists z with $f(z) < 0$ such that $v_i(z) = e_i f(z) < 0$. Hence $v_i(z)/e_i = \mathfrak{f}(z)$ and consequently $v_i(z)/e_i \leq v_j(z)/e_j$ if $j \neq i$. Hence if N is large, and we replace y by yz^N, we have in addition to $v_i(y)/e_i < v_j(y)/e_j$ if $j \neq i$ the required condition that $v_i(y)/e_i < 0$.

DEFINITION. *The valuations* $v_1,\ldots,v_k$ *of Theorem* 4.16 *will be termed the valuations associated with the filtration* f.

2. Miscellaneous results.

In this section we collect together a number of results which complement the valuation theorem. Our first result depends essentially on the fact that $\mathfrak{f}(x)$ takes rational values.

THEOREM 4.21. *Let* A *be a noetherian ring,* f *be a noether filtration on* A *and let* x *be an element of* A *such that* $\mathfrak{f}(x) < \infty$. *Then there exists a positive constant* C *depending on* x *but not on* n *such that*

$$n\mathfrak{f}(x) \geq f(x^n) \geq n\mathfrak{f}(x) - C.$$

The first inequality is immediate. To prove the second we note that $\mathfrak{f}(x)$ is a rational number q/m, where m is taken to be positive. Then $f^*(x^m) = q = \mathfrak{f}(x^m)$. It follows from Lemma 2.31 that there exists a constant k such that, for all r,

$$f(x^{mr}) \geq qr - k = mr\mathfrak{f}(x) - k.$$

Now let $n = rm + s$ with $0 \leq s < m$. Then

$$f(x^n) \geq f(x^{mr}) + f(x^s) \geq mr\mathfrak{f}(x) - k + f(x^s) \geq n\mathfrak{f}(x) - (k + s\mathfrak{f}(x) - f(x^s)),$$

and we take $C = \text{Max}(k + s\mathfrak{f}(x) - f(x^s))$ taken over $s = 0,\dots,m-1$.

For simplicity the next theorem is stated and proved only for noetherian domains.

THEOREM 4.22. *Let* A *be a noetherian domain, field of fractions* F, *and let* B *be a second noetherian domain such that* $F \supseteq B \supseteq A$. *Let* f *be a noether filtration on* A, g *be a noether filtration on* B *and* v *be a valuation associated with* f *which takes values other than* $0,\infty$. *Suppose further that,*

i) $g(x) \geq f(x)$ *on* A,

ii) $v(x) \geq 0$ *on the ring* $B_0 = B_0(g)$ *consisting of elements* $x \in B$ *such that* $g(x) \geq 0$,

iii) $v(g^+) = v(f^+)$; $v(f^-) = v(g^-)$.

Then v *is associated with* g.

Condition i) implies that G(g) contains G(f). Now the associated valuations v_i of f are the restrictions of the valuations $V_i(x)/m_i$ to A, where $V_i(x)$ ranges over the associated valuations of the principal filtration f_u on G(f). This follows from Lemma 4.13 and Theorem 4.16. Further, the valuations V_i are graded valuations by Lemma 4.12. Now we consider the condition on v_i that it takes values different from

0 or ∞. This implies that v_i is associated with either f^+ or f^- by the corollaries to Theorem 4.16. We take these in turn. Suppose the former. Then, by Theorem 4.16, $V_i(u)/m_i = v(f^+) = v(g^+)$. This, together with ii), implies that $V_i(x) \geq 0$ on $G(g^+)$. Hence, Corollary i) to Theorem 3.21 implies that V_i is a Krull valuation of $G(g)$, with $V_i(u) > 0$. Hence v as defined in Theorem 4.16 is associated with g^+ and hence with g. The case when v_i is associated with f^- is almost identical, replacing f^+ by f^-.

THEOREM 4.23. *Let* A *be a noetherian ring, and let* f *be a noether filtration on* A. *Let* S *be a multiplicatively closed sub-set of* A_0. *Then with the notation of section 4 of chapter 2,*

i) $f_S^*(x) \geq n$ *if and only if* $v_i(x) \geq ne_i$ *for those valuations* v_i *associated with* f *whose centres do not meet* S.

ii) $\mathbf{f}_S(x) = \mathrm{Min}\, v_i(x)/e_i$ *taken over the same set of* v_i *as in i).*

Lemma 2.41 implies that $f_S^*(x) \geq n$ if and only if $f^*(sx) \geq n$ for some s in S, and further, we can replace the element s by any of its powers. Hence $f_S^*(x) \geq n$ if and only if, for some s $v_i(sx) \geq ne_i$ for all associated valuations of f.

But, replacing s by a sufficiently high power, this condition will be satisfied, whatever x, for those valuations whose centres meet S, while, for the other valuations, we have $v_i(sx) = v_i(x)$ since $v_i(s) = 0$. Hence i) follows.

Next i) implies ii). Since the numbers e_i are all rational, it follows that there exists an integer m independent of x such that $v_i(x^m)/e_i$ is an integer for each v_i in i). Then $f_S^*(x^m) = \mathrm{Min}\,(v_i(x^m)/e_i)$. Hence

$$\mathbf{f}_S(x) = \mathbf{f}_S(x^m)/m \geq f_S^*(x^m)/m = \mathrm{Min}\,(v_i(x)/e_i) \geq f_S^*(x) \geq f_S(x),$$

and the result follows since $\mathrm{Min}\,(v_i(x)/e_i)$ is homogeneous.

THEOREM 4.24. *Let* A *be a noetherian domain,* f *be a noether filtration on* A. *Then there exists a noether filtration* f'(x) *on* A *such that*

i) $f'(x) \geq f(x)$ *for all* x, *and* f'(x) *is equivalent to* f(x),

ii) *if* $V_1,\ldots,V_k$ *are the Krull valuations of* G(f') *such that* $V_i(u) > 0$ *and* $\mathbf{p}_i$ *denotes the centre of* V_i *on* G(f') *then* $\mathbf{p}_i$ *has height* 1, *and if* $i \neq j$ *then* $\mathbf{p}_i \neq \mathbf{p}_j$.

We have $A[t,u] = G(f)_{(u)}$. Now $G(f)^*$ is graded. Hence by the Corollary iv) to Theorem 3.21, we can adjoin to G(f) elements of A[t,u], which may be taken to be homogeneous, and which are integrally dependent on G(f), such that the ring G' they generate is a finite G(f)-module and the centres of the valuations V such that $V(u) > 0$ have height 1. Clearly this ring determines a noether filtration f' on A such that $f'(x) \geq f(x)$ for all x in A and since $G' = G(f')$ is a finite G(f)-module, there exists an integer k such that $u^{-k}G(f) \supseteq G'$, i.e., $f(x) + k \geq f'(x)$ for all x, whence f, f' are equivalent.

5. THE STRONG VALUATION THEOREM

1. Preliminaries.

From now on, we restrict attention to noether filtrations which take only non-negative values, and all noether filtrations mentioned will be assumed to satisfy this condition. It follows that the symbol $v(f^-)$ will never occur, and we will therefore write $v(f)$ in place of $v(f^+)$.

We now consider the following question. Suppose that A is a noetherian ring and that f is a noether filtration on A. Then it is natural to ask whether the integral closure f^* of f is equivalent to f. Since $f^*(x) \geq f(x)$ for all x, this is equivalent to the statement that there exists a constant K such that $f(x) \leq f^*(x) \leq f(x) + K$ for all x. An equivalent formulation is that $u^{-K}G(f) \supseteq G(f^*)$, which in turn is equivalent to $G(f^*)$ being a finite $G(f)$-module, and hence implies that f^* is also a noether filtration.

Note that the restriction to non-negative noether filtrations implies that f and f^* are equivalent if and only if f^* is a noether filtration. For in this case $A_0(f^*) = A_0(f) = A$, and f^* is a noether filtration if and only if $G(f^*)$ is finitely generated over A. But this is equivalent to $G(f^*)$ being finitely generated over $G(f)$, and since $G(f^*)$ is an integral extension of $G(f)$, this in turn is equivalent to $G(f^*)$ being a finite $G(f)$-module and hence equivalent to f.

The existence of noether filtrations on A which are integrally closed in the sense that $f = f^*$ can impose restrictions on A. These restrictions relate to the completion of A with respect to a topology defined by f. We therefore interpose an account of the theory of completions, the basic results being given without proof. This account will occupy the next section.

2. Completions, the Cohen Structure Theorems and Nagata's Theorem.

Let A be a commutative ring and let f be a non-negative filtration on A, which, for simplicity, we assume to take integer values. Then we can define a topology on A in terms of f by taking as a basis of neighbourhoods of an element x the cosets $x+I_n(f)$. It is clear that the condition that two filtrations f, g should define the same topology is that, given n, there should exist integers n', n" such that $I_n(f) \supseteq I_{n'}(g)$ and $I_n(g) \supseteq I_{n''}(f)$. Hence, in particular, equivalent filtrations define the same topology. If we restrict attention to noether filtrations, then we can pin-point the exact condition that two noether filtrations define the same topology. Recall that the radical of f is the common radical of the ideals $I_n(f)$.

THEOREM 5.21. *If* A *is a noetherian ring, and* f, g *are two non-negative noether filtrations on* A *both of which take values other than* 0, ∞, *then* f, g *define the same topology if and only if they have the same radical. Further, if this is the case, then there exist constants* C,C' *such that* $g(x) \leq C.f(x)$ *and* $f(x) \leq C'.g(x)$ *for all* x *in* A.

Suppose that f, g do not have the same radical. Then either radf contains an element not in radg, or radg contains an element not in radf. Suppose the former. Choose an element x in radf but not in radg. Then x^r is not contained in $I_n(g)$ for any positive n, but for any given n, x^r is contained in $I_n(f)$ for all sufficiently large r. It follows that if n > 0, we cannot find n" such that $I_n(g) \supseteq I_{n''}(f)$, and hence the topologies are not the same. The case of the second alternative is similar.

Now suppose that f, g have the same radical J. Then we can define a noether filtration $f_J(x)$ by defining $f_J(x) \geq n$ if $x \in J^n$ and $f_J(x) = \infty$ if $x \in J^n$ for all n. It is clear that f_J has radical J. Now to prove that f,g define the same topology it is clearly sufficient to prove the last sentence for f,g and it will suffice to do this for the pairs f, f_J and g, f_J. We consider the first case only. First we note that there exists an integer p such that $I_1(f) \supseteq J^p$ from which it follows that $I_n(f) \supseteq J^{pn}$. Hence $f(x) \geq f_J(x)/p$. To prove the converse we need the precise description of a noether filtration. We recall that we can find elements $a_1,\ldots,a_s$ in J and weights $w(1),\ldots,w(s)$ which are positive integers such that $I_n(f)$ is generated by the monomials in $a_1,\ldots,a_s$ of weight ≥n. Now suppose that the maximum of the weights w(i) is m. Then the generators of $I_{mn}(f)$ are products of at least n a_i's and hence $J^n \supseteq I_{mn}(f)$. Hence $f_J(x) \geq f(x)/m$. This completes the proof.

We will refer to f_J as the J-*adic filtration* on A.

Now we return to the general case. If we suppose a filtration f given, we say that a sequence $\{x_r\}$ of elements of A is an f-*null sequence* if, given n, we can find an integer $r' = r'(n)$ such that $x_r \in I_n(f)$ if $r > r'(n)$. We term a sequence $\{x_r\}$ a *Cauchy sequence* if the sequence $\{x_r - x_{r+1}\}$ is a null sequence. If we now define addition, subtraction and multiplication of sequences by the rules, $\{x_r\} \pm \{y_r\} = \{x_r \pm y_r\}$ and $\{x_r\}\{y_r\} = \{x_r y_r\}$, then it is comparatively easy to prove that the set of Cauchy sequences forms a ring C and that the set of null sequences is an ideal N of this ring.

The quotient ring C/N will be termed the *completion* of A with respect to the topology defined by f and will be denoted by $A^\wedge(f)$. In fact, the notions of null and Cauchy sequences, and hence $A^\wedge(f)$, depend only on the topology defined by f, and in the case of noether filtrations on the radical J of f. In this case we will write $A^\wedge(J)$ in place of $A^\wedge(f)$ if this proves more convenient. We have a homomorphism h of A into $A^\wedge(f)$ in which an element x of A is mapped into the image in $A^\wedge(f)$ of the Cauchy sequence $\{x_r\}$ with $x_r = x$ for all r. The kernel of this homomorphism h is the intersection of the ideals $I_n(f)$. Further we can "lift" the filtration f to $A^\wedge(f)$, obtaining a filtration $f^\wedge$ on $A^\wedge(f)$ which is defined as follows. Suppose that an element x of $A^\wedge(f)$ is represented by a Cauchy sequence $\{x_r\}$. Then either $f(x_r)$ is constant for large n, and this constant value is taken as $f^\wedge(x)$, or $f(x_r) \to \infty$ as $r \to \infty$, and we take $f^\wedge(x) = \infty$. It is clear that $f^\wedge(h(x)) = f(x)$ for all x in A.

We now term a ring A *complete* with respect to a filtration f on it if every Cauchy sequence $\{x_r\}$ (with respect to f) of elements of A has a *unique* limit x (x being a limit of $\{x_r\}$ if the sequence $\{x - x_r\}$ is a null sequence). The uniqueness is equivalent to the assumption that $f(x) = \infty$ implies $x = 0$. We note at this point that to prove that A is complete with respect to a filtration f it is not necessary to prove that every Cauchy sequence has a limit. It is sufficient to prove that sequences $\{x_r\}$ which satisfy the condition that $f(x_r - x_{r+1}) \geq r$ for all r have a limit. Such a sequence will be referred to as a *special* Cauchy sequence. This follows since, given any Cauchy sequence $\{x_n\}$, we can always find a special Cauchy sequence $\{y_n\}$ such that $\{x_n - y_n\}$ is a null sequence. We omit the details. Then the ring $A^\wedge(f)$ is complete with respect to $f^\wedge$. Further, the ring $A^\wedge(f)$ and the homomorphism $h: A \to A^\wedge(f)$ already defined have the following property. If B is a ring complete with respect to a filtration g, and k is a homomorphism of $(A,f) \to (B,g)$ which is continuous in the sense that if $\{x_r\}$ is an f-null sequence then $\{k(x_r)\}$ is a g-null sequence, then there is a unique continuous homomorphism $k' : (A^{\wedge}(f), f^{\wedge}) \to (B,g)$ such that $k = k'h$.

The next lemma indicates the consequence of the assumption that f is integrally closed.

LEMMA 5.22. *If A is a commutative ring and f is an integrally closed filtration on A, then $A^{\wedge}(f)$ has no nilpotent elements other than zero .*

It is sufficient to show that if an element x of $A^{\wedge}(f)$ satisfies $x^2 = 0$, then $x = 0$. Suppose that x is the image of the Cauchy sequence $\{x_n\}$. Then, as $x^2 = 0$, the sequence $\{x_r^2\}$ is a null sequence. Hence, given an integer m, there exists an integer $n(m)$ such that if $r > n(m)$, $f(x_r^{\,2}) > 2m$. But, this implies that $f^*(x_r) \geq m$ and hence that $f(x_r) \geq m$. Hence $\{x_r\}$ is a null-sequence and consequently $x = 0$.

We now restrict attention to the case where A is a noetherian local ring with maximal ideal $\mathfrak{m}$ and assume f to satisfy the condition that $I_n(f)$ is $\mathfrak{m}$-primary for all n (it is sufficient that $I_1(f)$ be $\mathfrak{m}$-primary). Since we are only concerned with questions relating to the structure of $A^{\wedge}(f)$, it is sufficient for most purposes to replace f by $f_{\mathfrak{m}}$. We will therefore write $A^{\wedge}$ in place of $A^{\wedge}(f)$.

Now suppose that A is complete with respect to f, that is, the homomorphism $h: A \to A^{\wedge}$ is an isomorphism. Then we have a very powerful set of theorems collectively known as the Cohen Structure Theorems. These are, in fact, true for non-noetherian rings, but we will not make use of this fact. The basic theorem requires a definition. Let K be a field. We now associate with K a ring W(K), termed the *Witt* ring of K. If K has characteristic 0, W(K) is K itself. If K has characteristic p, then W(K) has the following properties.

i)It is a local domain of characteristic zero with one non-zero prime ideal pW(K) and W(K)/pW(K) = K,

ii) W(K) is complete with respect to the pW(K)-adic topology.

It is a fact, which we will not prove, that there does exist such a ring, and further, it is unique to within isomorphism.

The first of the Cohen Structure Theorems then states that if A is complete with respect to the $\mathfrak{m}$-adic topology, the isomorphism of K with $A/\mathfrak{m}$ can be lifted to a continuous homomorphism of W(K) into A.

An almost immediate consequence of this is that, if $x_1,\ldots,x_d$ are elements of $\mathfrak{m}$, then there is a unique extension of the homomorphism $W(K) \to A$ to a continuous homomorphism of the ring of formal power series $W(K)[[X_1,\ldots,X_d]]$ in indeterminates $X_1,\ldots,X_d$ into A, with X_i mapping into x_i. Further, if the elements $x_1,\ldots,x_d$ form a basis of $\mathfrak{m}$, this homomorphism is onto. One consequence of this is that, if K has characteristic zero, then A is a homomorphic image of $K[[X_1,\ldots,X_d]]$. This is true also if A and K have the same prime characteristic $p > 0$. It will be convenient to state the results of this paragraph as a theorem, for future reference.

THEOREM 5.23. *Let Q be a complete local ring, with maximal ideal* $\mathfrak{m}$, *residue field* K. *Let* $x_1,\ldots,x_n$ *be a basis of* $\mathfrak{m}$. *Then*

i) *if* Q *is equicharacteristic, i.e.,* K, Q *have the same characteristic, then* $Q = K[[X_1,\ldots,X_n]]/I$ *for some ideal* I, *the images of* $X_1,\ldots,X_n$ *being* $x_1,\ldots,x_n$;

ii) *if* Q *has characteristic* 0 *and* K *characteristic* $p > 0$, *then* $Q = W(K)[[X_1,\ldots,X_n]]/I$;

iii) *if* Q *has characteristic* p^r, $r > 1$, *and* K *has characteristic* p, *where* $p > 0$, *then* $Q = W(K)[[X_1,\ldots,X_n]]/I$, *where* I *contains* p^r.

Next suppose that $x_1,\ldots,x_d$ generate an $\mathfrak{m}$-primary ideal. Then A is a finite module over the image of $W(K)[[X_1,\ldots,X_d]]$ (which we will denote by $W(K)[[x_1,\ldots,x_d]]$ if the map of W(K) into A is an isomorphism into and by $K[[x_1,\ldots,x_d]]$ if the image of W(K) is isomorphic to K). In particular, if A has no nilpotent elements, so that A has characteristic 0 or a finite prime p, and the height of $\mathfrak{m}$ is d, then

i) if A and K have the same characteristic, $K[[x_1,\ldots,x_d]]$ is isomorphic to $K[[X_1,\ldots,X_d]]$, while,

ii) if A has characteristic 0 and K has characteristic $p > 0$, then if $x_1 = p$, $W(K)[[x_2,\ldots,x_d]]$ is isomorphic to $W(K)[[X_2,\ldots,X_d]]$.

We will not prove the Cohen Structure Theorems, but refer the reader to the original paper of Cohen (Cohen [1946]) or to [BCA] Chapter 9 for a recent account.

We now come to the theorem of Nagata referred to in the heading of this section. We recall that a local ring A is said to be *analytically unramified* if its

$\mathfrak{m}$-adic completion A^ has no nilpotent elements. Nagata's Theorem can then be stated as follows.

THEOREM 5.24. *Let* Q *be a complete local ring with no non-zero nilpotent elements and let* $\mathbf{p}_1,\dots,\mathbf{p}_s$ *be the minimal prime ideals of* Q. *Let* F *be the complete ring of fractions of* Q *(so that* F *is the direct sum of the fields of fractions* $F_1,\dots,F_s$ *of* $Q/\mathbf{p}_1,\dots,Q/\mathbf{p}_s$*). Then the integral closure of* Q *in* F *is a finite* Q*-module.*

We will not prove this theorem, but refer the reader to Nagata's proof in [LR] Theorem (32.1) on p.112, or to [M] Corollary 2 on p.234. Both these proofs are limited to the case where Q is a complete domain, but it is easy to see that there is a finite integral extension of Q isomorphic to the direct sum of the domains $Q/\mathbf{p}_i$ whence the result is immediate for complete local rings without nilpotent elements. We will however prove a corollary of Nagata's Theorem (also due to Nagata). First we require a simple lemma.

LEMMA 5.25. *Let* $(Q,\mathfrak{m},k,d)$ *be a local ring and let* Q^ *be the completion of* Q *with respect to the* $\mathfrak{m}$*-adic topology. Then,*

i) *if* J *is any ideal of* Q, $JQ\hat{}\cap Q = J$,

ii) *if* x *is not a zero divisor of* Q, *it is not a zero divisor of* Q^.

i) If x belongs to $JQ\hat{}\cap Q$, then there exists a sequence of elements $\{x_n\}$ of J such that $x-x_n \in \mathfrak{m}^n$. Hence x belongs to $\cap(J+\mathfrak{m}^n)$. But by Krull's intersection theorem applied to the local ring Q/J, this intersection is J.

ii) Suppose that y is an element of Q^ such that xy = 0. Then we can find a sequence of elements y_n of Q such that $y-y_n \in \mathfrak{m}^n Q\hat{}$, and hence xy_n belongs to $\mathfrak{m}^n$. But, as x is not a zero divisor of Q, the Artin-Rees Lemma implies the existence of an integer k such that $y_n \in \mathfrak{m}^{n-k}$ for all n. Hence $\{y_n\}$ is a null sequence and y = 0.

THEOREM 5.26. *Let* $(Q,\mathfrak{m},k,d)$ *be an analytically unramified local ring with complete ring of fractions* R. *Then the integral closure of* Q *in* R *is a finite* Q*-module.*

Let S be the set of non-zero divisors of Q. Then the elements of S are also non-zero divisors of Q^. Hence by Nagata's Theorem above, the integral closure of Q^ in $Q\hat{}_S$ is a finite Q^-module. It follows that there exists an element u of S such that, for all elements x of R integrally dependent on Q (and hence on $Q^{\wedge}$), ux belongs to $Q^{\wedge}$. But ux = y/z with y in Q and z in S. Hence $y \in zQ^{\wedge} \cap Q = zQ$ and therefore ux belongs to Q, i.e. the integral closure of Q in R is contained in $u^{-1}Q$ and so is a finite Q-module.

We conclude this section with a further observation concerning local rings and their $\mathfrak{m}$-adic completions. We have seen that if I is any ideal of a local ring Q, then $IQ^{\wedge} \cap Q = I$. Now suppose that f is a noether filtration on Q taking only non-negative values, and define $f^{\wedge}$ to be the extension of f to $Q^{\wedge}$ defined by the sequence of ideals $I_n(f)Q^{\wedge}$, then it follows that f is the restriction of $f^{\wedge}$ to Q. Further, as $f^{\wedge}$ is defined by the sequence of ideals $I_n(f)Q^{\wedge}$, it is also a noether filtration. The final observation we will need is that f^* is the restriction of $(f^{\wedge})^*$ to Q. This is most easily seen as follows. We have $(f^{\wedge})^*(x) \geq m$ if there exists a constant k such that $f^{\wedge}(x^n) \geq mn - k$ for all n. If x is in Q, this implies that $f(x^n) \geq mn - k$ for all n and hence that $f^*(x) \geq m$.

3. The Strong Valuation Theorem.

In this section we are concerned, to begin with, with the proof of the Strong Valuation Theorem for local rings. This states most simply that, if Q is a local ring which is analytically unramified and f is any noether filtration on Q which takes only non-negative values, then there is a constant K such that

$$0 \leq f^*(x) - f(x) \leq K.$$

The observations at the end of the last paragraph enable us to restrict attention to the case where Q is complete with no nilpotent elements.

Our main tool is a ring S(f) which we now define. We suppose that Q is a complete local ring and that f is a noether filtration on Q which we suppose is defined by a set of elements $a_1,\dots,a_s$ with weights $w(1),\dots,w(s)$, all >0. We now consider first the ring Q[[t]] of formal power series in an indeterminate t. This ring has no nilpotent elements if Q has no nilpotent elements. We now take S(f) to be the sub-ring of Q[[t]] consisting of all formal power series in the elements $a_1t^{w(1)},\dots,a_st^{w(s)}$. This is a sub-ring of Q[[t]] and can also be described as the ring of formal power series $\Sigma c_r t^r$ with $f(c_r) \geq r$.

LEMMA 5.31. S(f) *is a complete local ring.*

We have a homomorphism of the complete local ring $Q[[Y_1,\dots,Y_s]]$ onto S(f) in which a power series $g(Y_1,\dots,Y_s)$ over Q is mapped into the element $g(a_1t^{w(1)},\dots,a_st^{w(s)})$ of S(f); see [Z-S] Vol. 2, pp. 135-6. Hence S(f), as a homomorphic image of a complete local ring is itself a complete local ring. (The author owes this short proof to R.Y. Sharp. It replaces a much longer one.)

Now we are in a position to prove the strong valuation theorem for local rings.

THEOREM 5.32. *If* Q *is a local ring, with maximal ideal* $\mathfrak{m}$, *which is analytically unramified and* f *is a noether filtration on* Q *which takes only non-negative values, then there exists a constant* K(f) *depending only on* f, *such that*

$$f(x) \le f^*(x) \le f(x) + K(f).$$

We first impose a restriction on f, namely that there is a non-zero divisor w of Q such that f(w) > 0. Next we note that we can replace Q by its completion Q^ with respect to the $\mathfrak{m}$-adic topology and f by its extension to Q^. Hence we may assume that Q is complete with respect to the $\mathfrak{m}$-adic topology and that it contains no non-zero nilpotent elements. Hence the same is true of the ring S considered above. Now we can apply Nagata's Theorem to note that the integral closure of S in any sub-ring of its complete ring of fractions is a finite S-module. But as w is a non-zero divisor of Q and hence of S, it follows that t = wt/w is in the complete ring of fractions of S and hence the integral closure S' of S in S[t] is a finite S-module. Let $u_1(t),\dots,u_m(t)$ be a basis of the S-module S' and let K be the maximum of the degrees of the elements $u_i(t)$ considered as polynomials over S. Then it follows that each element of S' is a formal power series Σc_nt^n satisfying the condition that $f(c_r) \ge r - K$ for all r. Now suppose that x is an element of Q such that $f^*(x) \ge n$. Then xt^n is in S'. Hence $f(x) \ge n - K$, implying that $f^*(x) - f(x) \le K$, and we can take K(f) = K.

Now we remove the condition that there exists an element w which is not a zero divisor such that f(w) > 0. Adjoin an indeterminate X to Q and consider the filtration F(g(X)) on Q[X] defined by

$$F(c_0 + c_1X + \dots + c_nX^n) = \text{Min}\,(f(c_r) + r).$$

We note

a) the restriction of F to Q is f,

b) if h(X) is a polynomial such that the constant term of h(X) is a unit a of Q, then $F(g(X)h(X)) = F(g(X))$ for all polynomials $g(X)$.

To prove b), we first note that, if we write $h(x) = a(1 - Xk(X))$, and $q_N(X) = (1 - X^N(k(X))^N)/(1 - Xk(X))$, so that $q_N(X)$ has coefficients in Q, then

$$h(X)q_N(X) = a(1 - X^N(k(X))^N).$$

Hence, if N is large, $F(g(X)h(X)q_N(X)) = F(g(X))$ and, as the l.h.s. is $\geq F(g(X)h(X))$, which in turn is $\geq F(g(X))$, statement b) follows.

It follows that we can extend F to the local ring Q_X obtained by localising Q[X] at the maximal ideal $(\mathfrak{m},X)$ by defining $F(g(X)/h(X)) = F(g(X))$.

We now replace f,Q by F,Q_X, noting that

i) F is a noether filtration,

ii) Q_X is analytically unramified, its completion being $Q\hat{}[[X]]$,

iii) $F(X) > 0$ and X is not a zero divisor.

Thus the above proof works for F and we get the result for f by restriction to Q.

We now come to the global version of the Strong Valuation Theorem.

THEOREM 5.33. *Let A be a noetherian ring. Then the following conditions on A are equivalent.*

i) *For all noether filtrations* f *on* A, *there exists a constant* k(f) *such that*

$$0 \leq f^*(x) - f(x) \leq k(f)$$

for all non-zero x *in* A.

ii) *For every maximal ideal* $\mathfrak{m}$ *of* A, *the local ring* $A_\mathfrak{m}$ *is analytically unramified.*

First we prove that i) implies ii). Let $\mathfrak{m}$ be a maximal ideal of A and let f be the filtration on A defined by $f(x) = n$ if $x \in \mathfrak{m}^n - \mathfrak{m}^{n+1}$. If condition i) holds, then f and the integrally closed filtration f* induce the same topology, namely the $\mathfrak{m}$-adic topology on A. Then by Lemma 5.22, the $\mathfrak{m}$-adic completion of A contains no nilpotent elements. Since $\mathfrak{m}$ is maximal, an element x of A not in $\mathfrak{m}$ has an inverse in A^(f), i.e., the natural map of A into A^(f) extends to a map of $A_\mathfrak{m}$ into A^(f) and we have an isomorphism between the $\mathfrak{m}$-adic completion of A and the $\mathfrak{m}A_\mathfrak{m}$-adic completion of $A_\mathfrak{m}$, i.e., the latter is analytically unramified.

Next, ii) implies i). Suppose that $A_\mathfrak{m}$ is analytically unramified for all maximal ideals $\mathfrak{m}$ which contain radf. The set of prime ideals $\mathfrak{p}_1,\ldots,\mathfrak{p}_h$ associated with $I_n(f)$ for some n, is finite and hence we can choose a finite set of maximal ideals $\mathfrak{m}_1,\ldots,\mathfrak{m}_h$ of A such that $\mathfrak{m}_i$ contains $\mathfrak{p}_i$. Now let f_i denote the filtration on the localisation A_i of A at $\mathfrak{m}_i$ defined by the sequence of ideals $I_n(f)A_i$. Then clearly,

$$f(x) = \operatorname{Min} f_i(x)$$

on A, and if we interpret $f^*(x)$ as $[\mathfrak{f}(x)]$, it follows that

$$f^*(x) = \operatorname{Min} f_i^*(x).$$

But, by hypothesis, there exist constants $k(f_i)$ such that

$$f_i^*(x) - f_i(x) \leq k(f_i),\ i = 1,\ldots,h.$$

for all non-zero x of A_i. Hence

$$f^*(x) - f(x) \leq \operatorname{Max} k(f_i).$$

4. A criterion for analytic unramification.

In this section we use the Strong Valuation Theorem to prove a necessary and sufficient condition for a local ring Q to be analytically unramfied.

THEOREM 5.41. *Let Q be a local ring without nilpotent elements, and let R be its complete ring of fractions. Then the following two conditions on Q are equivalent.*

i) Q *is analytically unramified.*

ii) *For every finitely generated ring extension* $A = Q[x_1,\ldots,x_r]$ *of* Q *contained in* R, *the integral closure* A^* *of* A *in* R *is a finite* A*-module.*

i) implies ii). We can write $x_i = a_i/a_0$ with $a_0,\ldots,a_r$ in Q, a_0 being a non-zero divisor. Let J be the ideal generated by $a_0,\ldots,a_r$. Then every element of A is expressible in the form $a_0^{-m}y$, with $y \in J^m$ for m sufficiently large. Now suppose that z is an element of A^*. Then z satisfies an equation

$$z^n + d_1 z^{n-1} + \ldots + d_n = 0$$

with d_i in A. Then we can choose m such that $a_0^{im} d_i = c_i$ belongs to J^{im} for each i, and $w = a_0^m z$ satisfies

$$w^n + c_1 w^{n-1} + \ldots + c_n = 0.$$

Hence w is integrally dependent on Q. But, by the corollary to Nagata's Theorem

(Theorem 5.26), the integral closure Q^* of Q in R is a finite Q-module and therefore we can find a non-zero divisor u of Q such that $u^{-1}Q \supseteq Q^*$. Hence uw belongs to Q and satisfies an equation

$$(uw)^n + uc_1(uw)^{n-1} + \dots + u^n c_n = 0.$$

and since $u^i c_i$ belongs to J^{mi}, it follows that uw belongs to $(J^m)^*$. But, since Q is analytically unramified, the Strong Valuation Theorem implies that there exists an integer t such that $J^{m-t} \supseteq (J^m)^*$. Hence

$$z = a_0^{-t}u^{-1}(a_0^{-(m-t)}uw) \in a_0^{-t}u^{-1}A.$$

Therefore A^* is contained in $a_0^{-t}u^{-1}A$ and so is a finite A-module.

ii) implies i). Let $J = (a_0,\dots,a_r)$ be an $\mathfrak{m}$-primary ideal of Q, where $\mathfrak{m}$ is the maximal ideal of Q, a_0 being a non-zero divisor. Let A be the ring $Q[x_1,\dots,x_r]$, where $x_i = a_i/a_0$. Then, if y is an element of $(J^m)^*$, it is clear that y/a_0^m is contained in the integral closure A' of A in $A[a_0^{-1}]$, which by ii) is a finite A-module. Hence there exists an integer t such that A' is contained in $a_0^{-t}A$. This implies that $J^{m-t} \supseteq (J^m)^*$ for all m, which implies that f_J and f_J^* are equivalent, and hence that f_J^* is a noether filtration. But now, Lemma 5.22 implies that Q is analytically unramified.

COROLLARY. *If* Q *is an analytically unramified local ring, complete ring of fractions* R, A *is a finitely generated extension of* Q *contained in* R, $\mathfrak{p}$ *is a prime ideal of* A *and* $Q' = A_{\mathfrak{p}}$, *then* Q' *is analytically unramified.*

Let R' be the complete ring of fractions of Q', so that R' is a homomorphic image of R. Let $x_1',\dots,x_r'$ be elements of R', which are the images of elements $x_1,\dots,x_r$ of R. Then we can construct the ring $Q'[x_1',\dots,x_r']$ by first constructing the ring $A' = A[x_1,\dots,x_r]$ and then taking the ring of fractions with denominators the elements of A not in $\mathfrak{p}$. By the theorem above, the integral closure of A' in R is a finite A'-module. Hence the integral closure of $Q'[x_1',\dots,x_r']$ in R' is a finite $Q'[x_1',\dots,x_r']$-module. It now follows from the above that Q' is analytically unramified.

6. IDEAL VALUATIONS (1)

1. Introduction.

In this section we again restrict attention to noether filtrations which take only non-negative values on a ring A, take some finite positive value, thus satisfying $f(1) = 0$, and we will be concerned with the valuations v which are associated with such filtrations. We will refer to such valuations as the *ideal valuations* of A. To begin with we will consider some theorems which enable us to place restrictions on the filtrations we consider and on the ring A. We first deal with the latter. If A has minimal prime ideals $\mathbf{p}_1,\ldots,\mathbf{p}_h$, then we have already seen in Theorem 4.16 that the valuations associated with a noether filtration f on A are obtained by lifting the valuations associated with the filtrations $f/\mathbf{p}_j$, which are valuations on the fields of fractions of $A/\mathbf{p}_j$, to A. This may give rise to a degenerate valuation if radf is contained in $\mathbf{p}_j$. Excluding the degenerate valuations, the valuations associated with noether filtrations on A arise from those associated with noether filtrations on the noetherian domains $A/\mathbf{p}_j$. Hence, where convenient, we will restrict A to be a noetherian domain, but, where such a restriction is made, it will be stated explicitly.

LEMMA 6.11. *Let* f *be a noether filtration on a noetherian ring* A, *satisfying the restrictions indicated above, and let* g *be a second such noether filtration equivalent to* f, *such that* $g(x) \le f(x)$ *for all* x *with a standard set of generators* $x_1,\ldots,x_m$ *having weights* $w(1),\ldots,w(m)$. *Let* w *be the least common multiple of* $w(1),\ldots,w(m)$, *and let* J *be the ideal* $I_w(g)$. *Then, if* $f_J(x)$ *is the noether filtration defined by* $f_J(x) \ge n$ *if* $x \in J^n$, *there exists an integer* k *such that, if* $n = qw + r$ *with* $q \ge k$ *and* $0 \le r < w$,

$$I_n(f) = J^{q-k}.I_{kw+r}(f),$$

and f *is equivalent to* $w.f_J$.

The elements $x_i t^{w(i)}$ of G(f) generate an ideal which contains all elements of sufficiently large positive degree, since f,g are equivalent. Hence if $s(i) = w/w(i)$, the elements $(x_i t^{w(i)})^{s(i)} = x_i^{s(i)} t^w$ also generate such an ideal. But the elements $x_i^{s(i)}$ belong to J. Hence the ideal $Jt^wG(f)$ also contains all elements of sufficiently large positive degree. It follows that there exists an integer k such that if $n \geq kw$,

$$I_{n+w}(f) = J.I_n(f),$$

and the first statement of the lemma is immediate. It implies that, if $f(x) = n = qw + r$ ($q \geq k$ and $0 \leq r < w$), then $x \in J^{q-k}$ and hence

$$wf_J(x) \geq w(q-k) > f(x) - (k+1)w.$$

Also, if $f_J(x) = m$, $x \in J^m$ and so belongs to $I_{mw}(f)$. Hence $f(x) \geq wf_J(x)$. Hence f, $w.f_J$ are equivalent.

COROLLARY. *If* μ *is a multiple of* kw, *then*

$$I_{n\mu}(f) = I_\mu(f)^n.$$

It is sufficient to take $\mu = kw$, since, if $\mu = vkw$, and the result is true for kw in place of μ, then

$$I_{n\mu}(f) = I_{nvkw}(f) = (I_{kw})^{vn} = (I_{kw}{}^v)^n = I_\mu{}^n.$$

Since $I_{kw} \supseteq J^k$, we have $I_{nkw} = (J^k)^{n-1}.I_{kw}.(I_{kw})^n, I_{nkw}$, and the result is proved.

Now it is clear that the valuations associated with $w.f_J$ and f_J are the same. Hence the valuations associated with f and f_J are the same. In the first series of papers (Rees[1955],[1956]a,b,c) in which ideal valuations appeared, the only filtrations considered were the filtrations of the type f_J, and the valuations associated with f_J were referred to as the valuations associated with the ideal J. It is for this reason that they are referred to as ideal valuations in this set of notes.

The next class of noether filtrations we shall consider are the basic filtrations.

DEFINITIONS. A *filtration* g *equivalent to* f *is said to be a reduction of* f *if* $g(x) \leq f(x)$ *for all* x.

A noether filtration f *will be said to be basic if it has a (standard) set of* r *generators and no reduction g of* f *can be generated by fewer than* r *elements.*

As remarked in chapter 2 after the definition of sets of generators of a noether filtration, the restriction to standard sets makes no difference to this definition. Also we refer to sets of generators of f rather than f^+, since, by our blanket restriction in this chapter, $f = f^+$.

(Note; the condition that g is a reduction of f can be replaced by the condition that g is equivalent to f without altering this definition. This will be proved later.)

At this point it is convenient to relate these definitions to the earlier theory of reductions of ideals developed by D.G. Northcott and the author in Northcott and Rees[1954a] and [1954b].

We recall briefly some of the definitions. If J is an ideal of a noetherian ring A, and K is an ideal contained in J, we term K a *reduction* of J if, for some r,

$$J^{r+1} = J^r K.$$

Since this implies that $J^n \supseteq K^n \supseteq J^r K^n = J^{n+r}$ for all $n \geq 0$, it follows that f_K is a reduction of f_J. Conversely, if f_K is a reduction of f_J, then, applying Lemma 6.1 to f_J, f_K, so that $w = 1$ and $r = 0$ with the notation of that lemma, we obtain

$$J^n = J^{n-k} K^k$$

for some k and $n > k$, proving that K^k is a reduction of J^k.

In [1954b] the notion of a generalised reduction of J was introduced. This can be described as a set of elements $x_1,\ldots,x_r$ of J such that for suitable positive integers $w(1),\ldots,w(r)$ and n large enough,

$$J^n = x_1 J^{n-w(1)} + \ldots + x_r J^{n-w(r)}.$$

This is equivalent to the statement that, if g is the filtration generated by $x_1,\ldots,x_r$ with weights $w(1),\ldots,w(r)$, then g is a reduction of f_J.

Finally, an ideal is termed *basic* if its only reduction is itself. However, f_J may have reductions different from itself even in this case (see the example considered in chapter 2 at the end of section 4), and hence the definition of basic filtration takes the slightly different form given above.

We will also need another group of definitions generalising the definition of the form rings of an ideal.

DEFINITIONS. *If f is a noether filtration on a noetherian ring* A, *we will refer to the ring* G(f)/uG(f) *as the form ring of* f *and denote it by* F(f).

If J *is any ideal of* A, *then the ring* G(f)/uG(f) + JG(f) = F(f)/JF(f) *will be denoted by* F(f,J).

The notion of basic filtration is particularly important in the case where A is a local ring. We therefore now consider this case, and take A to be a local ring Q with maximal ideal $\mathfrak{m}$ and residue field k = Q/$\mathfrak{m}$. The Krull dimension of Q will be taken to be d. We will often abbreviate this by writing A = (Q,$\mathfrak{m}$,k,d).

We now require one more definition. This uses the definition of the spread of a graded ring as defined in chapter 1, section 4 (prior to Theorem 1.43).

DEFINITION. *The spread of a noether filtration* f *on* A = (Q,$\mathfrak{m}$,k,d) *is the spread of the graded ring* F(f,$\mathfrak{m}$), *and will be denoted by* s(f).

We recall at this point the two alternative definitions of s(M) for a finitely generated G-module where G is a positively graded, finitely generated extension of a field G_0 = k. The first one is the general one, that s(M) is the smallest integer s such that there exists an irrelevant ideal of G/AnnM generated by s elements. The second is that denoted in chapter 1 by s(a,M) and is defined as the least integer s such that $n^{-s}.a(M_n) \to 0$ as $n \to \infty$, where $a(M_n) = \dim_k M_n$. Note that the second definition implies that, if M' is a sub-module of M, then s(M') ≤ s(M).

THEOREM 6.12. *Let* A = (Q,$\mathfrak{m}$,k,d).

i) *Let* f *be a noether filtration on* A *with* s(f) = s *and* b *be a basic reduction of* f. *Let* $x_1, \dots, x_n$ *be a set of generators of* b *containing as few elements as possible, and let* w(i) *be the weight of* x_i. *Then* n = s, *and, if* X_i *is the image of* $x_i t^{w(i)}$ *in* G(f,$\mathfrak{m}$), $X_1, \dots, X_s$ *are algebraically independent over* k.

ii) *If* b *is a basic noether filtration on* A, *then* F(b,$\mathfrak{m}$) *is isomorphic to* $k[X_1, \dots, X_s]$, *where* s = s(b), *and* $X_1, \dots, X_s$ *are independent homogeneous indeterminates over* k *of appropriate degrees.*

iii) *If* f,g *are equivalent noether filtrations,* s(f) = s(g).

iv) s(f) *equals the minimal number of generators of a filtration equivalent to* f.

v) $s(f_J) = s(w.f_J)$ *for any ideal* J *and any positive integer* w.

Before proceeding to the proof, we require a lemma.

LEMMA 6.121. $s(f) = s(G^+(f))$, *where* $G^+(f)$ *is the graded ring consisting of all sums of elements of* G(f) *of degree* ≥0.

First we observe that $s(G^+(f)) = s(G^+(f)/\mathfrak{m}G^+(f))$ by Theorem 1.43. Write H for the graded ring $G^+(f)/\mathfrak{m}G^+(f)$ and F for $F(f,\mathfrak{m})$. Then F is a homomorphic image of H, the kernel of the homomorphism being $(uG(f)\cap G^+(f)) + \mathfrak{m}G^+(f)/\mathfrak{m}G^+(f)$. Suppose that X is a homogeneous element of $G^+(f)$ whose image in H is contained in this kernel. Since $I_1(f) \neq Q$, it follows that if X has degree 0, X is contained in $\mathfrak{m}G^+(f)$, and so its image is 0. Suppose that X has degree $r > 0$. Let k,w be as in Lemma 6.11. Then X^{kw} belongs to $u^{kw}G(f)$ and so X^{kw} is of the form zt^{rkw} with z in $I_{(r+1)kw}$. But by Lemma 6.11, $I_{(r+1)kw} = I_{rkw}(I_w)^k$, and so X^{kw} is contained in $\mathfrak{m}G^+(f)$. Hence the image of X in H is nilpotent. Since the kernel of the map H to F is contained in the radical of H, it now follows that s(H) = s(F), which proves our result.

Now we proceed to the proof of Theorem 6.12.

i)Let B denote the image of $F(b,\mathfrak{m})$ in $F(f,\mathfrak{m})$. As b,f are equivalent, $F(f,\mathfrak{m})$ is a finite B-module. It follows that $s(f) = s(f,\mathfrak{m}) \leq s(B)$, and is equal to s(B) since B is a sub-ring of $F(f,\mathfrak{m})$ and so a sub-module over B. On the other hand, $s = s(B) \leq n$ with equality if $X_1,\ldots,X_n$ are algebraically independent. Suppose they are not. Then there exist $s < n$ homogeneous elements $Y_1,\ldots,Y_s$ of B such that B and hence $F(f,\mathfrak{m})$ is a finite $k[Y_1,\ldots,Y_s]$-module, or, equivalently, such that $Y_1,\ldots,Y_s$ generate an irrelevant ideal of $F(f,\mathfrak{m})$. Let Y_i be the image of $z_i = y_i t^{w'(i)}$ under the map $G^+(f) \to F(f,\mathfrak{m})$, where z_i belongs to $G^+(b)$. Since the images of $z_1,\ldots,z_s$ in $F(f,\mathfrak{m})$ generate an irrelevant ideal of $F(f,\mathfrak{m})$, and the kernel of the map $G^+(f)/\mathfrak{m}G^+(f)$ onto $F(f,\mathfrak{m})$ is nilpotent, it follows that the images of $z_1,\ldots,z_s$ in $G^+(f)$ also generate an irrelevant ideal of $G^+(f)/\mathfrak{m}G^+(f)$. Then it follows that, for large N,

$$(z_1G^+(f) +\ldots+ z_sG^+(f))_N + \mathfrak{m}G^+(f)_N = G^+(f)_N,$$

whence by Nakayama's Lemma, $z_1,\ldots,z_s$ generate an irrelevant ideal of $G^+(f)$. It now follows that the elements y_i, $i = 1,\ldots,s$, if given weights $w'(i)$, $i = 1,\ldots,s$, generate a noether filtration g such that g is equivalent to f and hence to b, and $g(x) \leq b(x)$ for all x so that g is a reduction of b. Hence $n > s$. But as b is basic, the definition of n implies that $n = s$ and hence that $X_1,\ldots,X_s$ are algebraically independent over k.

ii) This follows from i) by taking $f = b$.

iii) Let f be generated by $x_1,\ldots,x_m$ with weights $w(1),\ldots,w(m)$, and let g be generated by $y_1,\ldots,y_n$ with weights $w'(1),\ldots,w'(n)$. Then the noether filtration h generated by the elements x_i with weight $w(i)$, $i = 1,\ldots,m$, and y_j with weight $w'(j)$, $j = 1,\ldots,n$, is equivalent to both f and g and $h(x) \geq f(x),g(x)$ for all x. Hence it is sufficient to take the case where $g(x) \geq f(x)$ for all x. Now there exists a basic filtration b such that $b(x) \leq f(x)$ for all x and b is generated by $s(f)$ elements and not less. But $b(x) \leq g(x)$ and b is equivalent to g. Hence the minimal number of generators of b is also equal to $s(g)$.

iv) This follows from the fact that if f is generated by n elements then $s(f) \leq n$ and i) above, since there certainly exist basic filtrations b equivalent to f satisfying $b(x) \leq f(x)$ for all x.

v) Consider the rings $F(f_J,\mathfrak{m})$ and $F(w.f_J,\mathfrak{m})$. The set of elements of degree n in the former consists of elements xt^n with x belonging to $J^n/J^n\mathfrak{m}$. In the latter, the set of elements of degree n is zero save when n is a multiple of w. For, in $G(w.f_J)$, the set of elements of degree $n = qw + r$, $0 < r \leq w$, is the module $J^{q+1}t^n$, implying that $uG(w.f_J)$ contains all homogeneous elements of positive degree not divisible by w. Further, the set of elements of $F(w.f_J,\mathfrak{m})$ of degree qw is the set of elements xt^{qw} with x in $J^q/J^q\mathfrak{m}$. Hence the ring $F(w.f_J,\mathfrak{m})$ is derived from the ring $F(f_J,\mathfrak{m})$ by multiplying degrees by w. Using the definition of $s = s(a,M)$ as the least integer s such that $n^{-s}\dim_k M_n$ tends to zero, we see at once that $s(F(f_J,\mathfrak{m})) = s(F(w.f_J,\mathfrak{m}))$ which proves the result.

COROLLARY. *If* f *is as in the theorem there exists an ideal* J *generated by* s *elements where* $s = s(f)$, *such that* f *is equivalent to* $w.f_J$ *for some integer* w.

By definition, we can choose a filtration g which is a reduction of f generated by elements $x_1,\dots,x_s$, with weights $w(1),\dots w(s)$. Let w be the least common multiple of $w(1),\dots,w(s)$, and let $r(j) = w/w(j)$. Then the filtration generated by the elements $y_j = x_j^{w(j)}$ each with weight w, is a reduction of g and hence of f. But if J is the ideal $(y_1,\dots,y_s)$, this filtration is $w.f_J$.

Note that Theorem 6.12, when applied to f_J, implies the existence of generalised reductions $x_1,\dots,x_s$ of J, where $s = s(J)$ is the analytic spread of J. However, to prove the existence of *reductions* of J generated by s elements it is necessary to assume that k is infinite. (See Northcott & Rees[1954a]).

DEFINITION. *Let* f *be a noether filtration on a noetherian ring* A. *Then the height and altitude of* f *are, respectively, the minimum and maximum of the heights of the associated prime ideals of* radf.

In the lemma following, and in many places thereafter, we will make use of the altitude or dimension inequality. This can be stated as follows.

THE ALTITUDE INEQUALITY. *Let* A *be a noetherian domain, with field of fractions* F, *and let* B *be a second noetherian domain finitely generated over* A, *with field of fractions* E. *Let* $\mathfrak{P}$ *be a prime ideal of* B *meeting* A *in* $\mathfrak{p}$, *and let* K,k *denote the fields* $B_{\mathfrak{P}}/\mathfrak{P}B_{\mathfrak{P}}$ *and* $A_{\mathfrak{p}}/\mathfrak{p}A_{\mathfrak{p}}$ *respectively. Then*

$$\mathrm{ht}\mathfrak{p} + \mathrm{trans.deg.}_F E \geq \mathrm{ht}\mathfrak{P} + \mathrm{trans.deg.}_k K.$$

A proof can be found in [Z-S], vol. 2, appendix 1 as propositions 1 and 2, the latter being stated in the above form on p.326, and is termed the *dimension inequality* (note that Zariski & Samuel use dimension to mean transcendence degree).

LEMMA 6.13. *Let* f *be a non-negative filtration on* $A = (Q,\mathfrak{m},k,d)$. *Then*

$$d \geq s(f) \geq \mathrm{alt}\, f.$$

Choose J as in Theorem 6.12, Corollary. Then $s(f_J) = s(f)$, and J,f have the same radical, implying that alt J = alt f. Hence it is sufficient to prove the inequalities with f_J replacing f. Since J is generated by s = s(f) elements, the second inequality now follows from Krull's altitude theorem which implies that $s \geq \text{alt}\, J$ (= alt f).

Writing f in place of f_J, we now consider the ring $F(f,\mathfrak{m})$ which is isomorphic to $G(f)/(uG(f) + \mathfrak{m}G(f))$ and also to $k[X_1,\dots,X_s]$ where s = s(f), and $X_1,\dots,X_s$ are algebraically independent over k. Hence $uG(f) + \mathfrak{m}G(f)$ is a graded prime ideal $\mathfrak{P}$ of G(f). Let Q' be the localisation of G(f) at $\mathfrak{P}$. Then the residue field of Q' is a pure transcendental extension of k of transcendence degree s. Now let $\mathfrak{P}'$ be a minimal prime ideal of G(f) contained in $\mathfrak{P}$, so that $\mathfrak{P}'$ is graded. This will meet Q in a prime ideal $\mathfrak{p}$ of Q.

Next we observe that the field of fractions of $Q'/\mathfrak{P}'Q'$ is the same as that of $G(f)/\mathfrak{P}'$ and so has transcendence degree 1 over that of $Q/\mathfrak{p}$, since $G(f)/\mathfrak{P}'$ is a sub-ring of $(Q/\mathfrak{p})[t,u]$ containing $u = t^{-1}$. The altitude inequality now gives us

$$\dim Q'/\mathfrak{P}'Q' + s \leq \dim Q/\mathfrak{p} + 1.$$

But $\dim Q'/\mathfrak{P}'Q' \geq 1$. Hence $s \leq \dim Q/\mathfrak{p} \leq d$.

DEFINITION. *Let* v *be a valuation on a noetherian ring* A, ≥0 *on* A *and with centre* $\mathfrak{p}$. *Let* K_v *be the residue field of* v *and let* $k = A_{\mathfrak{p}}/\mathfrak{p}A_{\mathfrak{p}}$. *Then we define the dimension of* v *over* A, *written* $\dim_A v$, *to be the transcendence degree of* K_v *over* k.

LEMMA 6.14. *Let* f *be a noether filtration on* $A = (Q,\mathfrak{m},k,d)$ *and let* v *be a valuation associated with* f, *the centre of* v *being* $\mathfrak{m}$. *Then*

$$\dim_Q v \leq s(f) - 1 \leq d - 1.$$

As in the last lemma, we can replace f by f_J where J is the ideal of the corollary to Theorem 6.12 (note that the valuations associated with $f, w.f_J$ are the same since $f, w.f_J$ have the same integral closure, while $w.f_J$ and f_J have the same associated valuations since, if $f = w.f_J$, $\mathfrak{f} = w.\mathfrak{f}_J$ and we can apply Lemma 2.12). Now consider the ring G(f) and the graded Krull valuation V on G(f) of which a multiple of v is a restriction. The centre $\mathfrak{P}$ of V on G(f) contains $uG(f) + \mathfrak{m}G(f)$. Hence $K_{\mathfrak{P}} = G(f)_{\mathfrak{P}}/PG(f)_{\mathfrak{P}}$ has transcendence degree at most s over k, and since V is a Krull valuation of G(f), K_V is a finite algebraic extension of $K_{\mathfrak{p}}$. Hence K_V has transcendence degree at most s over k. But K_V has transcendence degree 1 over K_v and hence the first inequality follows. The second follows from 6.13.

For convenience we now state a particular case of Theorem 4.22 which we need. Recall that, as we are only dealing with filtrations which take no negative values in this chapter, we write v(f) for what has earlier been denoted by $v(f^+)$.

THEOREM 6.15. *Let* f,g *be two noether filtrations on a noetherian ring* A *such that* g(x)≥f(x) *for all* x *and let* v *be a valuation which takes values other than* 0,∞. *Suppose further that a)* v *is associated with* f *and b)* v(g) = v(f). *Then* v *is associated with* g.

Let $\mathfrak{p}$ be the minimal prime ideal of A on which v(x) = ∞. Then v is associated with the filtration $f|\mathfrak{p}$ on $A/\mathfrak{p}$, and it will be sufficient to prove that v is associated with $g|\mathfrak{p}$. Hence we may assume that A is a domain. Since f,g are assumed not to take negative values, it follows that $v(f^-)= v(g^-) = 0$. Hence the three conditions of 4.22 are satisfied and the result follows.

COROLLARY. *Let* f *be a noether filtration on* A = (Q,$\mathfrak{m}$,k,d) *and let* v *be a valuation associated with* f *which has centre* $\mathfrak{m}$. *Then there exists a noether filtration* g *on* Q *with radical* $\mathfrak{m}$ *such that* g(x) ≥ f(x) *for all* x, v(g) = v(f), *and* v *is associated with* g.

Since the first two statements imply the third, we only have to find g with radical $\mathfrak{m}$ satisfying the first two conditions. Let J be any $\mathfrak{m}$-primary ideal such that v(J) > v(f), where v(J) is the minimum of v(x) on J. Such ideals exist, since v has centre $\mathfrak{m}$. For we only have to take J contained in $\mathfrak{m}^r$, where $rv(\mathfrak{m}) > v(f)$. We now take g to be $f \cup f_J$ where this is the filtration obtained by adding to the generators of f the elements of a basis of J, each taken with weight 1. Since the radical of a filtration contains the ideal generated by a set of generators of the filtration, the radical of g contains J and must be $\mathfrak{m}$. It is obvious that g(x) ≥ f(x) for all x and, finally, the choice of J ensures that v(g) = v(f).

Now we turn in the remaining sections of this chapter to the problem of characterising the ideal valuations of a noetherian domain A. We have already remarked that it is sufficient to consider only domains, since the ideal valuations of A are determined by those of the domains $A/\mathfrak{p}$ where $\mathfrak{p}$ runs over the minimal prime ideals of A. Next, if v is associated with a noether filtration f on A and $\mathfrak{p}$ contains the centre of v on A, then v is associated with the noether filtration f_S on the local ring $A_\mathfrak{p}$, where $S = A - \mathfrak{p}$. Hence the set of ideal valuations of A is the union of the sets of ideal valuations of $A_\mathfrak{p}$, where $\mathfrak{p}$ runs over all maximal ideals of A (alternatively, it is union of the sets of ideal valuations of $A_\mathfrak{p}$ with centre $\mathfrak{p}A_\mathfrak{p}$ where $\mathfrak{p}$ ranges over all non-zero prime ideals of A). Hence it is sufficient to consider only the case of local domains. This we do in the remaining sections.

2. The ideal valuations of a local domain.

We commence this section by recalling briefly the essential properties of A-sequences, grade and Cohen-Macaulay local rings. Proofs are not always provided. Where this is the case, the reader is referred to [Z-S] vol. 2, Appendix 6 "Macaulay Rings", and all references below are to this source. There are certain differences between our notation and terminology and that of [Z-S], and this is indicated below as appropriate. Finally, note that we do not assume that the rings A,... referred to below are domains except where this is stated.

DEFINITION. *A sequence of elements* $x_1,\dots,x_r$ *of a ring* A *is called an* A*-sequence if*

$$(x_1A + x_2A + \dots + x_{i-1}A) : x_i = x_1A + x_2A + \dots + x_{i-1}A$$

for $i = 1,\dots,r$, *that is, for each* i, *the image of* x_i *in* $A/(x_1,\dots,x_{i-1})$ *is not a zero divisor.*

If J *is an ideal of* A, *then the grade of* J *is defined to be the length of the longest* A*-sequence contained in* J. *This will be denoted by* grade J.

In [Z-S] the term *prime sequence* is used in place of A-sequence. The term *grade* is not defined in this generality, but only when J is the maximal ideal $\mathfrak{m}$ of a local ring Q, and in this case, what we term grade $\mathfrak{m}$ above is termed the grade of Q. They also use the term *homological codimension* and use the abbrevation codh Q. In this context we will use grade Q.

We note that the definition of grade given above is consistent with the definition given in chapter 3 of the phrases "grade J = 1 and grade J > 1".

Now suppose $(Q,\mathfrak{m},k,d)$ is a local ring.

DEFINITION. Q *is said to be Cohen-Macaulay* (C-M) *if* grade Q = dimQ = d.

We now list a number of results on A-sequences, grade and C-M local rings we shall need. These will be listed as 6.2 followed by a letter.

6.2A. If J is an ideal generated by an A-sequence $(a_1,\dots,a_n)$, and $\mathfrak{p}$ is any associated prime ideal of A associated with J, then ht$\mathfrak{p} \geq n$, with equality if and only if $\mathfrak{p}$ is minimal over J.

Proof: see [Z-S] p.394 and the references to [Z-S] vol. 1 given there.

Let A = $(Q,\mathfrak{m},k,d)$ and $(a_1,\dots,a_n)$ be a Q-sequence contained in $\mathfrak{m}$, generating an ideal J. Then the following statements are true.

6.2B. Any permutation of this sequence is also a Q-sequence.

Proof: [Z-S] lemma 2 on p.395.

6.2C. The sequence $(a_1{}^r,\dots,a_n{}^r)$ is a Q-sequence.

First we note that, if a_n is prime to $(a_1,\dots,a_{n-1})$ so is $a_n{}^r$. Hence $a_1,\dots,a_{n-1},a_n{}^r$ is a Q-sequence. Permuting the sequence, we can similarly raise the elements $a_1,\dots,a_{n-1}$ to the rth power.

6.2D. A sequence of elements $a_1,\dots,a_n$ of Q is a Q-sequence if and only if the images of $a_1,\dots,a_n$ in the completion Q^ of Q is a Q^-sequence.

This is proved in the course of the proof of lemma 6 on p.400 of [Z-S].

6.2E. Q is Cohen-Macaulay, if and only if Q^ is Cohen-Macaulay.

This is lemma 6 on p.400 of [Z-S].

6.2F. If $(Q,k,\mathfrak{m},d)$ is Cohen-Macaulay, and $(x_1,\dots,x_r)$ is a sequence of elements of Q, $r \leq d$, then it is a Q-sequence if and only if it can be extended to a sequence $(x_1,\dots,x_d)$ which generates an $\mathfrak{m}$-primary ideal.

[Z-S], p.399, corollary 2.

6.2G. If $(Q,\mathfrak{m},k,d)$ is Cohen-Macaulay and $\mathfrak{p}$ is any prime ideal of Q, then

$$\mathrm{ht}\mathfrak{p} + \dim Q/\mathfrak{p} = d.$$

[Z-S], p.399, corollary 3.

LEMMA 6.21. *Let* $(Q,\mathfrak{m},k,d)$ *be a local ring and let* $(x_1,\dots,x_r)$ *be a Q-sequence generating an ideal J. Then:*

i) if $\mathfrak{p}$ *is a prime ideal of Q associated with the zero ideal,* $\dim(Q/\mathfrak{p}) \geq r$;

ii) if $f = f_J$, $F(f)$ *is isomorphic to* $A[X_1,\dots,X_r]$ *where* $A = Q/J$ *and the indeterminates* $X_1,\dots,X_r$ *are the images of* $x_1t,\dots,x_rt$ *under the homomorphism* $G(f) \to F(f)$.

i) We use induction on r, the result being trivial if $r = 0$. Hence we can assume that the result is true for the ring Q/x_1Q, that is, if $\mathfrak{P}$ is a prime ideal associated with x_1Q, then $\dim Q/\mathfrak{P} \geq r - 1$. But, by [Z-S] p.394 lemma 1, we can choose $\mathfrak{P}$ properly containing $\mathfrak{p}$, and hence $\dim Q/\mathfrak{p} > \dim Q/\mathfrak{P}$ which proves the result.

ii) To prove this result, we have to show that if $p(X_1,\dots,X_r)$ is a homogeneous polynomial of degree n over Q such that $p(x_1,\dots,x_r) \in J^{n+1}$, then the coefficients of p belong to J. Since any element of J^{n+1} can be expressed as a homogeneous polynomial of degree n with coefficients in J, it is sufficient to prove that if $p(X_1,\dots,X_r)$ satisfies the stronger condition that $p(x_1,\dots,x_r) = 0$ then its coefficients belong to J. This we prove by a double induction on r and n.

First we deal with the case $r = 1$ and all n. In this case, since x_1 is not a zero divisor, if $p(X_1) = aX_1{}^n$, the condition $ax_1{}^n = 0$ implies that $a = 0$ and hence certainly belongs to J.

Next we deal with the case $n = 1$, and use induction on r, the case $r = 1$ having been dealt with. Suppose that $p(X_1,\dots,X_r) = a_1X_1 + \dots + a_rX_r$. Then, if

$$a_1x_1 + \dots + a_rx_r = 0$$

the definition of a Q-sequence implies that a_r can be written in the form $b_1x_1 + \dots + b_{r-1}x_{r-1}$ (and so belongs to J), and hence we have

$$(a_1 + a_rb_1x_r)x_1 + \dots + (a_{r-1} + a_rb_{r-1}x_r)x_{r-1} = 0$$

and our inductive assumption now implies that, if $i<r$, $a_i+a_rb_ix_r \in J$, and hence $a_i \in J$.

We now suppose that both r and n are >1. Suppose that $p(X_1,\dots,X_r)$ satisfies the condition that $p(x_1,\dots,x_r) = 0$ and that p has degree n. Then we can write

$$p(X_1,\dots,X_r) = q(X_1,\dots,X_{r-1}) + X_r r(X_1,\dots,X_r)$$

where q has degree n and r has degree n-1. First suppose that $q(X_1,\dots,X_{r-1})$ is the zero polynomial. Then, as x_r is not a zero divisor, $r(x_1,\dots,x_r) = 0$ and our inductive hypothesis on n implies that the coefficients of r and hence of p belong to J. We now reduce the general case to this case. We replace Q by $Q/x_rQ = Q'$. Then if x_i' denotes the image of x_i in Q', $x_1',\dots,x_{r-1}'$ is a Q'-sequence. Further, if $q'(X_1,\dots,X_{r-1})$ is the polynomial over Q' obtained by reducing the coefficients of q modulo x_rQ, then $q'(x_1',\dots,x_{r-1}') = 0$. Hence, by our inductive hypothesis on r, the coefficients of q' belong to J/x_rQ and those of q belong to J. We can now apply the above case to the polynomial p - q. Note that we have used the fact that Q is local to imply that a permutation of a Q-sequence is a Q-sequence.

COROLLARY. If Q *is* C-M, *then* Q *is unmixed.*

For, by i), all prime ideals $\mathbf{p}$ associated with (0) in Q satisfy $\dim Q/\mathbf{p} = \dim Q$.

We now use this last result to prove the second part of the following.

THEOREM 6.22. i) *Let* $A = (Q,\mathfrak{m},k,d)$ *be a local domain and let* v *be an integer-valued valuation on* Q, ≥ 0 *on* Q *and* >0 *on* $\mathfrak{m}$. *Further, suppose that the residue field* K_v *of* v *has transcendence degree* d-1 *over* k. *Then* v *is an ideal valuation of* Q *and* K_v *is a finitely generated extension of* k.

ii) *If* $(Q,\mathfrak{m},k,d)$ *is a Cohen-Macaulay ring, or is a domain which is a homomorphic image of a Cohen-Macaulay ring, then all ideal valuations having centre* $\mathfrak{m}$ *on* Q *have dimension* d-1.

i). Let $x_1,\dots,x_{d-1}$ be elements of K_v which form a transcendence base of K_v over k. We now choose elements $a_1,\dots,a_d$ in Q such that x_i is the image of a_i/a_d in K_v. We can replace a_i by $a_i + b_i$ for $i = 1,\dots,d$, where $v(b_i) > v(a_i)$ (note that $v(a_1) = \dots = v(a_d)$, since $v(x_i) = 0$). It follows, by a standard prime-avoidance argument, that we can further assume that $a_1,\dots,a_d$ generate an $\mathfrak{m}$-primary ideal J. We now take $f = f_J$, and construct the ring G(f). We can extend v to a graded valuation V on G(f) by giving v(u) the value equal to the common value of $v(a_i)$, $i = 1,\dots,d$. Let $\mathfrak{P}$ be the centre of V on G(f) so that $\mathfrak{P} \cap Q = \mathfrak{m}$. It is clear that the images of ta_i, $i = 1,\dots,d$, in G(f) are algebraically independent over k. Hence applying the altitude inequality to Q and the localisation of G(f) at $\mathfrak{P}$, we obtain

$$d + 1 \geq \operatorname{ht} \mathfrak{P} + d$$

whence $\mathfrak{P}$ has height ≤ 1 and hence 1. It follows from Theorem 3.24 that V is a Krull valuation on G(f) and V(u) > 0. Hence v is associated with f (or with the ideal J) by Theorem 4.16.

ii) If v is an ideal valuation of a Cohen-Macaulay local ring $(Q,\mathfrak{m},k,d)$ with centre $\mathfrak{m}$, then v is associated with f_J for some $\mathfrak{m}$-primary ideal J. We can replace J by any power of J and also by any reduction. Since a suitable power of J has a reduction generated by d elements (see Northcott-Rees[1954b] or the Corollary to 6.12 above), we can assume that $J = (a_1,\dots,a_d)$, and $a_1,\dots,a_d$ is then a Q-sequence. Let $f = f_J$. Then, by the last lemma, F(f) is isomorphic to $(Q/J)[X_1,\dots,X_d]$ where $X_1,\dots,X_d$ are indeterminates. It follows that uG(f) is a primary ideal and so has only one associated prime $\mathfrak{P}$. Further, $G(f)/\mathfrak{P}$ is isomorphic to $k[X_1,\dots,X_d]$ where $X_1,\dots,X_d$ are indeterminates. Hence the graded valuation V on G(f) of which a multiple of v is the restriction has centre $\mathfrak{P}$ by Lemma 3.211, and so K_V has transcendence degree d over k. Hence K_v has transcendence degree $d - 1$ over k, i.e., $\dim v = d - 1$.

Finally we turn to the case of a local domain Q which is a homomorphic image of a Cohen-Macaulay local ring C. Suppose that $Q = C/\mathfrak{p}$. Then if C has dimension n and Q has dimension d, we can find elements $a_1,\dots,a_{n-d}$ in $\mathfrak{p}$ which form a C-sequence. Then $C/(x_1,\dots,x_{n-d})C$ is a Cohen-Macaulay local ring C' of dimension d having Q as a homomorphic image. It follows that $Q = C'/\mathfrak{p}'$, where $\mathfrak{p}'$ is a minimal prime ideal of C'. Hence every ideal valuation of Q lifts to an ideal valuation of C', and this lifting does not alter the residue field. It follows that the dimension of every ideal valuation of Q centre $\mathfrak{m}$ is $d-1$.

COROLLARY. *If* $(Q,\mathfrak{m},k,d)$ *is a complete local domain, then all ideal valuations on* Q *centre* $\mathfrak{m}$ *have dimension* $d-1$.

For every complete local domain is a homomorphic image of a ring $R = W(k)[[X_1,\dots,X_n]]$ of formal power series over $W(k)$ where $W(k)$ is either k or a complete valuation ring. In either case the generators of the maximal ideal form an R-sequence.

We have thus characterised completely the ideal valuations centre $\mathfrak{m}$ on a complete local domain $(Q,\mathfrak{m},k,d)$. They are those which have dimension $d-1$. We now use this to give our first characterisation of the ideal valuations centre $\mathfrak{m}$ on an arbitrary local domain.

THEOREM 6.23. *Let* $(Q,\mathfrak{m},k,d)$ *be a local domain and let* f *be a noether filtration on* Q *whose radical is* $\mathfrak{m}$. *Let* $f^\wedge$ *be the extension of* f *to the completion* $Q^\wedge$ of Q *defined by the sequence of ideals* $I_n(f)Q^\wedge$. *Then the associated valuations of* $f^\wedge$ *are the extensions by continuity* $v^\wedge_1,\dots,v^\wedge_q$ *of the valuations* $v_1,\dots,v_q$ *associated with* f. *The restriction of* $v^\wedge_i$ *to* Q *is* v_i *and* v_i, $v^\wedge_i$ *have the same dimension.*

Since $I_n(f)Q^\wedge \cap Q = I_n(f)$ for all n, it follows that f, $\mathfrak{f}$ are the restrictions of $f^\wedge$, $\mathfrak{f}^\wedge$ respectively to Q. Hence, if

$$\mathfrak{f}^\wedge(x) = \operatorname{Min} v^\wedge_i(x)/v^\wedge_i(f^\wedge)$$

where i runs from 1 to q, is the irredundant representation of $\mathfrak{f}^\wedge$ as a lower bound of valuations,

$$\mathfrak{f}(x) = \operatorname{Min} v_i(x)/v_i(f^\wedge)$$

is a representation of $\mathfrak{f}$, where v_i is the restriction of $v^\wedge_i$ to Q (implying that $v^\wedge_i$ is the extension of v_i to $Q^\wedge$). This representation is also irredundant. For suppose x_i is an element of $Q^\wedge$ such that

$$v\hat{}_i(x_i)/v\hat{}_i(f\hat{}) < v\hat{}_j(x_i)/v\hat{}_j(f\hat{}) \quad \text{if } j \neq i,$$

then we can replace x_i by an element of Q differing from it by an element of a sufficiently high power of $\mathfrak{m}Q\hat{}$ without altering these inequalities. Hence $v_1,\dots,v_q$ are the associated valuations of f, and $v\hat{}_1,\dots,v\hat{}_q$ are their extensions by continuity.

We now consider the residue fields of $v, v\hat{}$, where v is a valuation ≥ 0 on Q and >0 on $\mathfrak{m}$, which takes integer values and $v\hat{}$ is its extension to $Q\hat{}$ by continuity. It is clear that K_v is a sub-field of $K_{v\hat{}}$. Now let $\mathfrak{p}$ be the prime ideal of $Q\hat{}$ on which $v\hat{}(x) = \infty$, so that $v\hat{}$ is a valuation on the field of fractions of $Q\hat{}/\mathfrak{p}$. Write Q'' for $Q\hat{}/\mathfrak{p}$. Now suppose that x is an element of $K_{v\hat{}}$. Then we can find y, z in Q'' such that x is the image of y/z. We can, in fact, choose y,z to be the images of elements of Q under the map of Q into Q'', since there exist such images differing from y,z by elements of $(\mathfrak{m}'')^n$, for arbitrarily large n. Hence the map of K_v into $K_{v\hat{}}$ is onto and the residue fields of $v, v\hat{}$ are isomorphic. Hence if v is an ideal valuation v, $v\hat{}$ have the same dimension.

We can now draw a number of conclusions from the above and the fact that an ideal valuation, centre the maximal ideal $\mathfrak{m}$ of Q, is associated with a noether filtration radical $\mathfrak{m}$ by the corollary to Theorem 6.15. First suppose that Q is a complete local ring with minimal prime ideals $\mathfrak{p}_1,\dots,\mathfrak{p}_h$, and that the dimension of the local ring $Q/\mathfrak{p}_j$ is d_j. Then the dimensions of the ideal valuations of Q, centre $\mathfrak{m}$, are contained in the set of integers $d_j - 1$, $j = 1,\dots,h$, and further, if f is any noether filtration with radical $\mathfrak{m}$, then the set of associated valuations $v_1,\dots,v_q$ contains valuations taking the value ∞ on $\mathfrak{p}_j$ for each j and therefore contains valuations with dimension each of the integers $d_1 - 1,\dots,d_h - 1$.

Now suppose that $(Q,\mathfrak{m},k,d)$ is any local ring. Then we can draw the same conclusions concerning the dimensions of the ideal valuations centre $\mathfrak{m}$, if we first replace Q by its completion and then take d_j to be the dimension of $Q\hat{}/\mathfrak{p}_j$ where $\mathfrak{p}_j$ is a minimal prime ideal of $Q\hat{}$. Further, we note that if v is an integer-valued valuation on Q with centre $\mathbf{m}$ and dimension $d_0 - 1$, then it is an ideal valuation if its extension to Q^ takes value ∞ on a minimal prime ideal $\mathbf{p}_0$ of Q^ such that $\dim Q\hat{}/\mathbf{p}_0 = d_0$.

We now introduce some terminology and a definition. Let A be a noetherian ring and $\mathbf{p}$ be a prime ideal of A. Let Q be the local ring $A_{\mathbf{p}}$ and let Q^ be its completion. Then we will say that an integer h_0 is a *quasi-height* of $\mathbf{p}$ if there exists a minimal prime ideal $\mathbf{p}_0$ of Q^ such that $Q\hat{}/\mathbf{p}_0$ has dimension h_0.

DEFINITION. *A local ring* $(Q,\mathbf{m},k,d)$ *is termed quasi-unmixed if* d *is the only quasi-height of* $\mathbf{m}$. *A noether ring* A *is termed quasi-unmixed if* $A_{\mathbf{p}}$ *is quasi-unmixed for all prime ideals* $\mathbf{p}$ *of* A.

We will now summarise the results of the discussion above in the following theorem.

THEOREM 6.24. *Let* A *be a noetherian ring,* $\mathbf{p}$ *be a prime ideal of* A, *and suppose that the set of quasi-heights of* $\mathbf{p}$ *is* $h_1,\dots,h_r$. *Then an ideal valuation* v *on* A *with centre* $\mathbf{p}$ *has dimension equal to* $h_j - 1$ *for some* j. *Further,*

i) *if* f *is a noether filtration on* A *such that* $\mathbf{p}$ *is a minimal prime ideal over* radf, *then the set of the associated valuations of* f *with centre* $\mathbf{p}$ *includes valuations with dimensions over* A *equal to* $h_j - 1$ *for* $j = 1,\dots,r$,

ii) *a valuation* v *on* A *with centre* $\mathbf{p}$ *and* $\dim_A v = h - 1$, *where* h *is the height of* $\mathbf{p}$ *(and hence the largest quasi-height), is an ideal valuation of* A,

iii) *if* A *is quasi-unmixed (i.e., every prime ideal has exactly one quasi-height), then the ideal valuations of* A, *centre* $\mathbf{p}$, *are precisely those satisfying*

$$\dim_A v = \mathrm{ht}\mathbf{p} - 1.$$

We conclude this chapter with a result closely related to a well-known result of L.J. Ratliff, (Ratliff[1974], Theorem 2.12 on p.189).

THEOREM 6.25. *Let* A *be quasi-unmixed and let* f *be a noether filtration on* A *such that* rad f *has height* h, *and, for every maximal ideal* $\mathfrak{m}$ *of* A *containing* rad f, *the filtration* $f_{\mathfrak{m}}$ *on the local ring* $A_{\mathfrak{m}}$ *has spread* h. *Then for all* n *the ideal* $I_n(f^*)$ *is unmixed of height* h.

By Lemma 6.13, alt$f_{\mathfrak{m}} \le h$ for all $\mathfrak{m}$, and, since rad f has height h, this implies that rad$f_{\mathfrak{m}}$ is unmixed of height h for all $\mathfrak{m}$ and hence that rad f is unmixed of height h. Hence we have to prove that, for all n, $I_n(f^*)$ has no imbedded primes. But this will be true if and only if $I_n(f_{\mathfrak{m}}^*)$, which equals $I_n(f^*)A_{\mathfrak{m}}$ by Lemma 2.41, is unmixed of height h for all $\mathfrak{m}$.

Hence it will be sufficient to prove the result when A is a local ring $(Q,\mathfrak{m},k,d)$, f has spread h, and rad f has height h. We may, by replacing f by an equivalent filtration, assume that f is generated by h elements $x_1,\ldots,x_h$. Now suppose that $\mathfrak{p}$ is a prime ideal associated with $I_n(f^*)$. But, by the Valuation Theorem, $I_n(f^*)$ is the intersection of the primary ideals $\mathfrak{q}_i$ defined as the set of elements x of Q such that $v_i(x) \ge nv_i(f)$, where v_i ranges over the valuations associated with f. Hence $\mathfrak{p}$ is the radical of one of the ideals $\mathfrak{q}_i$, i.e., is the centre of a valuation v associated with f. Since f is generated by h elements, $\dim_Q v \le h-1$, by Lemma 6.14, and hence ht$\mathfrak{p} \le h$. But rad f has height h, and hence $\mathfrak{p}$ must have height at least h. Hence the associated prime ideals of $I_n(f^*)$ all have height h, i.e., $I_n(f^*)$ is unmixed of height h.

7. IDEAL VALUATIONS (2)

1. Introduction.

The purpose of this chapter is to relate the set of ideal valuations of a finitely generated domain B over a noetherian domain A to the ideal valuations of A itself. For this purpose the description of the ideal valuations of A in terms of the completions of the localisations of A is not convenient and we therefore use a somewhat different one, namely that a valuation ≥ 0 on A is an ideal valuation of A if and only if there is a finitely generated extension B of A with the same field of fractions, such that $v(x) \geq 0$ on B and the centre of v on B has height 1. Note that this implies that v is a Krull valuation of B by Theorem 3.24, and we could weaken the above condition by simply requiring that v be a Krull valuation of B. The proof of this criterion is obtained by putting together Theorem 4.24 and the Corollary to Lemma 6.11. This is done in the proof of the following theorem.

THEOREM 7.11. *Let* A *be a noetherian domain,* v *be a valuation on the field of fractions* F *of* A *such that* $v(x) \geq 0$ *on* A. *Then* v *is an ideal valuation of* A *if and only if there exists a finitely generated extension* B *of* A, *contained in* F, *such that* $v(x) \geq 0$ *on* B *and the centre* $\mathbf{p}$ *of* v *on* B *has height* 1. *It is sufficient that* v *be a Krull valuation of* B.

Suppose that such an extension B exists. Then we can write $B = A[x_1/x_m,\dots,x_{m-1}/x_m]$, where $x_1,\dots,x_m$ belong to A, and we can assume that $v(x_m) > 0$. Hence $v(x_i) \geq v(x_m) > 0$. Now let $J = (x_1,\dots,x_m)A$ and $f = f_J$. Let V be the graded extension of v on the ring G(f) defined by taking $V(u) = v(x_m)$. Let $y = x_m t$ and consider the ring $G(f)[y^{-1}]$. It is clear that the elements of this ring of degree zero are the elements of B and that this ring is expressible in the form $B[y,y^{-1}]$. Further, $V(x) \geq 0$ on $B[y,y^{-1}]$ and $V(y) = 0$. Hence, if $\mathbf{p}$ is the centre of v on B, the centre of V on $B[y,y^{-1}]$ is $\mathbf{p}B[y,y^{-1}]$ and so has height 1. This implies that the centre of V on G(f) has height 1. Hence V is a Krull valuation of G(f), by Theorem 3.24, and satisfies $V(u) > 0$. Hence v is a valuation associated with f, by Theorem 4.16, and therefore is an ideal valuation of A.

Now suppose that v is associated with a noether filtration f. We can take this to be of the form f_J, where J is an ideal $(a_1,\dots,a_m)$, and further we can suppose that

$v(a_m) \leq v(a_i)$ for $i < m$. Then, if $C = A[a_1/a_m,\ldots,a_{m-1}/a_m]$, $v(x) \geq 0$ on C. Then, as above, if S is the multiplicatively closed set of powers of $a_m t$ in $G = G(f_J)$, then

$$G_S = C[X,X^{-1}]$$

where $X = a_m t$. Then, if we extend v to a graded valuation V on G, by taking $V(u) = v(a_m)$, V is associated with the filtration f_{uG} on G, since v is associated with f_J. Hence V is a Krull valuation of $C[X,X^{-1}]$. Now the integral closure of $C[X,X^{-1}]$ in its field of fractions is $C^*[X,X^{-1}]$, where C^* is the integral closure of C. Hence, by the Mori-Nagata Theorem, V has centre a height 1 prime ideal of $C^*[X,X^{-1}]$ which will be graded since V is graded. But, as X is a unit in $C^*[X,X^{-1}]$, any graded prime ideal $\mathfrak{P}$ of $C^*[X,X^{-1}]$ is of the form $\mathfrak{p}[X,X^{-1}]$, where $\mathfrak{p} = \mathfrak{P} \cap C^*$. Hence $\mathfrak{p}$ also has height 1 and is the centre of v, implying that v is a Krull valuation of C. Hence there exists a finite integral extension B of C such that the centre of v on B has height 1 by Theorem 3.24. This also proves the last sentence.

COROLLARY. *Let* $\mathfrak{p}$ *be a prime ideal of* A. *Then* h' *is a quasi-height of* $\mathfrak{p}$ *if and only if there is a finitely generated extension* B *of* A *and a height* 1 *prime ideal* $\mathfrak{P}$ *of* B *such that* $\mathfrak{P} \cap A = \mathfrak{p}$ *and* $B_{\mathfrak{P}}/\mathfrak{P}B_{\mathfrak{P}}$ *is an extension of* $A_{\mathfrak{p}}/\mathfrak{p}A_{\mathfrak{p}}$ *of transcendence degree* h'-1.

By Theorem 6.24, h' is a quasi-height of $\mathfrak{p}$ if and only if there exists an ideal valuation of A with centre $\mathfrak{p}$ whose residue field has transcendence degree h'-1 over $A_{\mathfrak{p}}/\mathfrak{p}A_{\mathfrak{p}}$. If there exists such a valuation we can choose B such that $v(x) \geq 0$ on B and the height of the centre $\mathfrak{P}$ of v on B has height 1. Hence, by the Krull-Akizuki Theorem, the residue field of v is algebraic over $B_{\mathfrak{P}}/\mathfrak{P}B_{\mathfrak{P}}$ and so the latter has transcendence degree h' - 1 over $A_{\mathfrak{p}}/\mathfrak{p}A_{\mathfrak{p}}$.

Conversely, if there exist B, $\mathfrak{P}$ as stated, then any valuation v centre $\mathfrak{P}$ is an ideal valuation of A by the theorem above. Since its residue field is an algebraic extension of $B_{\mathfrak{P}}/\mathfrak{P}B_{\mathfrak{P}}$, it follows that its residue field is an extension of $A_{\mathfrak{p}}/\mathfrak{p}A_{\mathfrak{p}}$ of transcendence degree h' - 1 and so h' is a quasi-height of $\mathfrak{p}$.

Before the next theorem we require two lemmas.

LEMMA 7.12. *Let* $(Q,\mathfrak{m},k)$ *be a 1-dimensional local domain field of fractions* F *and let*

E *be a finite algebraic extension of* F. *Then the set* Σ *of valuations* w *on* E *such that* $w(x) \geq 0$ on Q, >0 *on* $\mathfrak{m}$, *is finite and non-empty, and, for all* $w \in \Sigma$, K_w *is a finite algebraic extension of* k. *Finally, the integral closure of* Q *in* E *is the intersection of the rings* O_w, $w \in \Sigma$.

Let $u_1,\ldots,u_n$ be a basis of E over F. Then we can choose $a \neq 0$ in Q such that au_i is integral over Q for $i = 1,\ldots,n$. Let $L = Q[au_1,\ldots,au_n]$. Then L is a finite Q-module. Hence if x is a non-zero element of $\mathfrak{m}$, L/xL has finite length as a Q-module. It follows that L has only a finite number of non-zero prime ideals, and these are all maximal. Further, for each such maximal ideal $\mathfrak{n}$, the residue field $k_{\mathfrak{n}}$ of $L_{\mathfrak{n}}$ is a finite extension of k.

Hence the set Σ consists of all non-trivial valuations on E satisfying the condition that $w(x) \geq 0$ on L, and if $w \in \Sigma$, it has centre some maximal ideal $\mathfrak{n}$ of L and $O_w \supseteq L_{\mathfrak{n}}$ and hence $L_{\mathfrak{n}}*$. We now apply the Krull-Akizuki Theorem to $L_{\mathfrak{n}}$. Then $L_{\mathfrak{n}}*$ is contained in only a finite number of discrete valuation rings. Their intersection is $L_{\mathfrak{n}}*$ and their residue fields are finite extensions of $k_{\mathfrak{n}}$ and hence of k. Hence Σ is the union of a finite number of finite sets and so is finite. Further, as their centres are of height 1, the valuations w of Σ are the Krull valuations of L, and, by the Mori-Nagata Theorem, L* is the intersection of the rings O_w, $w \in \Sigma$. But L* is the integral closure of Q in E. Finally, Σ is not empty. For if it were, every element of E and hence of F, would be integrally dependent on Q. Let x be a non-unit of Q. Then if x^{-1} is integral on Q,

$$x^{-(r+1)} = a_0 + \ldots + a_r x^{-r},\ a_i \in Q$$

and hence, multiplying by x^r, $x^{-1} \in Q$, which is a contradiction.

LEMMA 7.13. *Let* A,B $\supseteq$ A *be two noetherian domains with fields of fractions* F *and a finite algebraic extension* E *of* F. *Let* $\mathfrak{P}$ *be a prime ideal of* B, *such that* $\mathfrak{p} = \mathfrak{P} \cap A$ *is of height* 1. *Then* $\mathfrak{P}$ *has height* 1.

By localising at $\mathfrak{p}$, $\mathfrak{P}$ we can assume that A, B are local with maximal ideals $\mathfrak{p}$, $\mathfrak{P}$ and that A has dimension 1. B contains a basis $u_1,\ldots,u_n$ of E over F, and, as in the last lemma, we can choose a in A so that $L = A[au_1,\dots,au_n]$ is a semi-local ring whose maximal ideals all have height 1. $\mathfrak{P}\cap L$ is a maximal ideal $\mathfrak{n}$ of L, which has height 1. Hence replacing A by $L_{\mathfrak{n}}$, we have reduced the proof to the case E = F, retaining the condition that A is local of dimension 1. We now assume that this is the case. Since A has dimension 1, T(A) = F. Hence B is noetherian, and, for any $x \neq 0$ in A B/xB is a finitely generated A/xA-module and so has finite length as an A/$\mathfrak{p}$-module, and hence as a B/$\mathfrak{P}$-module. It follows that B has dimension 1, since there is a principal $\mathfrak{P}$-primary ideal of B.

THEOREM 7.14. *Let* A, B $\supseteq$ A *be noether domains with fields of fractions* F,E, *where* E *is a finite algebraic extension of* F *and* w *be a valuation on* E *taking values* ≥ 0 *on* B, *with restriction* v *to* F, *then*

i) *if* v *is an ideal valuation of* A, w *is an ideal valuation of* B,

ii) *If* E = F, *so that* w = v, *and* B *is a finitely generated extension of* A *and* v *is an ideal valuation of* B, *then it is an ideal valuation of* A.

i) If v is an ideal valuation of A, then there is a finitely generated extension $A[x_1,\dots,x_m]$ of A on which v has centre of height 1. But then the centre of w on $B[x_1,\dots,x_m] \supseteq A[x_1,\dots,x_m]$ is also of height 1 and hence w is an ideal valuation of B.

ii) If v is an ideal valuation of B, then there exists a finitely generated extension $C = B[x_1,\dots,x_m]$ of B, contained in F, such that v has centre of height 1 on C. Since C is also finitely generated over A, v is also an ideal valuation of A.

DEFINITION. *An integral domain* D *with field of fractions* F *is termed a Krull domain if there exists a family of valuations* $\Sigma = \Sigma(D)$, *termed the essential valuations of D on* F *with the following properties:*

a) *all essential valuations are discrete rank* 1 *valuations;*

b) x *in* F *belongs to* D *if and only if* $v(x) \geq 0$ *for all* $v \in \Sigma$;

c) *if* $x \neq 0$ *belongs to* F, *then* $v(x) = 0$ *for all save a finite set of* $v \in \Sigma$;

d) *if* $v \in \Sigma$, *and* $\mathbf{p}(v)$ *denotes the centre of* v *on* D, *then* $O_v = D_{\mathbf{p}(v)}$, *implying that* $\mathbf{p}(v)$ *has height* 1.

Note that Theorem 3.21 implies that the integral closure A* of a noetherian domain in its field of fractions is a Krull domain, the essential valuations being the Krull valuations.

Condition d) serves to characterise the set of essential valuations among those sets which satisfy a),b),c). This is a consequence of the following lemma, taken from [Z-S] vol. 2, (theorem 26 on p.82).

LEMMA 7.15. *Let* Σ^* *be a set of valuations on the field of fractions* F *of a Krull domain* D *which satisfies a),b),c) in the above definition. Then* $\Sigma^* \supseteq \Sigma$.

Let $\mathbf{p}$ be a non-zero prime ideal of D. Then, if we show that some valuation v of Σ^* has centre contained in $\mathbf{p}$, the result will follow by taking $\mathbf{p} = \mathbf{p}(v)$ for v in Σ, since d) implies that the only non-trivial valuation with centre in $\mathbf{p}(v)$ is v.

Suppose the result false. Choose $x \neq 0$ in $\mathbf{p}$ and let $v_1,\ldots,v_s$ be the valuations in Σ^* for which $v_i(x) > 0$, the set being finite by c). If $\mathbf{p}_i$ is the centre of v_i, then by hypothesis, we can choose y_i in $\mathbf{p}_i$ but not in $\mathbf{p}$ for each i. Hence if $y = (y_1 \ldots y_s)^n$, where $n \geq \text{Max } v_i(x)$, then $v(y) \geq v(x)$ for all v in Σ^*. For this is true if v is one of the valuations v_i by our choice of y, and if v is any other valuation of Σ^*, $v(y) \geq 0 = v(x)$. Then b) implies that $z = y/x$ belongs to D. Hence $y = xz$ belongs to $\mathbf{p}$, contradicting the hypothesis.

LEMMA 7.16. *Let* A *be an integral domain integrally closed in its field of fractions* F *and let* f(X) *be a monic polynomial with coefficients in* A. *Then, if* f(X) = g(X)h(X), *where* g(X), h(X) *are monic polynomials with coefficients in* F, *the coefficients of* g(X),h(X) *belong to* A.

We can find a finite algebraic extension E of F in which f(X) splits into linear factors $(X-\alpha_1)\ldots(X-\alpha_N)$. Since f(X) is monic and has coefficients in A, the elements α_i of E are integral over A. The coefficients of g(X), h(X) are symmetric functions of sub-sets of the set $(\alpha_1,\ldots,\alpha_N)$ and so are also integral over A. Since these coefficients are in F, they belong to A.

THEOREM 7.17. *Let* A *be a noetherian domain with field of fractions* F, *and* A* *be the integral closure of* A *in a finite algebraic extension* E *of* F. *Then* A* *is a Krull domain, and every essential valuation of* A* *is an extension of a Krull valuation of* A *to* E.

If $u_1,\ldots,u_n$ is a basis of E over F, then for suitable a in A, au_i is integral over A for each i, and hence A* contains B = A[$au_1,\ldots,au_n$], and A* = B*. But B is a noetherian domain and has field of fractions E. Hence A* as the integral closure of a noetherian domain in its field of fractions is a Krull domain.

Take Σ^* to be the set of extensions of the Krull valuations of A to E. Applying Lemma 7.12 to the valuation ring O_v of a Krull valuation of A, we see that each Krull valuation of A has only a finite set of extensions to E and that these are discrete of rank.

Let x be any element of A*. Then x is a zero of a monic polynomial with coefficients in A. This polynomial is divisible by the minimum polynomial $m_x(X)$ of x over F (which is also a monic polynomial over F). Hence the coefficients of $m_x(X)$ all belong to the integral closure A' of A in F by the last lemma. Conversely, if the coefficients of the minimum polynomial $m_x(X)$ of an element x of E all belong to A', then x is integrally dependent on A' and hence on A. Now suppose the monic polynomial of x is

$$X^r + a_1X^{r-1} + \ldots + a_r.$$

Let v be a Krull valuation of A. Then we can characterise the integral closure O_v* of O_v in E in two ways. First, if x belongs to E, x belongs to O_v* if and only if the coefficients of $m_x(X)$ belong to O_v by the last paragraph applied to A = O_v. Secondly, by Lemma 7.12, $x \in O_v$* if and only if $w(x) \geq 0$ for all the extensions w of v to E. Applying the first characterisation to all v, we see that x belongs to O_v* for all v if and only if the coefficients of $m_x(X)$ belong to O_v for all v, i.e., belong to A', implying that $x \in$ A*. Hence A* is the intersection of the rings O_v*. Using the second characterisation, A* is the intersection of the rings O_w with $w \in \Sigma^*$. We now verify that Σ^* satisfies a),b),c). Condition a) is clearly satisfied. Condition b) was proved above. Hence we are left with c). It is sufficient to prove that if $x \neq 0$ is in A*, $w(x) > 0$ for only a finite set of w. Suppose $w(x) > 0$. Then, as $w(a_r) > 0$, w is an extension of one of the finite set of Krull valuations v of A satisfying $v(a_r) > 0$. Since each such v has only a finite number of extensions, it follows that $w(x) > 0$ for only a finite set of w in Σ. Hence all Krull valuations of A belong to Σ^* by Lemma 7.15.

In the next theorem, we make use of the fact that, if v is an ideal valuation of A with centre $\mathbf{p}$, and if $\dim_A v = r$, then $r + 1$ is a quasi-height of $\mathbf{p}$. This is proved in Theorem 6.24.

THEOREM 7.18. *Let* A *be a noetherian domain, field of fractions* F, *and let* v *be an ideal valuation on* F ≥ 0 *on* A, *centre* $\mathbf{p}$. *Then,*

i) *if* $Q = A_{\mathbf{p}}$ *has grade* r, $\dim_A v \geq r - 1$,

ii) v *is a Krull valuation of* A *if* $\dim_A v = 0$, *implying that the Krull valuations of* A *are the ideal valuations* v *of* A *satisfying* $\dim_A v = 0$.

i) For $Q^\wedge$ contains a $Q^\wedge$-sequence of length r. Hence, if $\mathfrak{P}$ is a prime ideal of $Q^\wedge$ associated with the zero ideal of $Q^\wedge$, in particular a minimal prime ideal, then $\dim Q/\mathfrak{P} \geq r$ by Lemma 6.21 i). Hence all quasi-heights of $\mathbf{p}$ are $\geq r$, and $\dim_A v \geq r - 1$.

ii) If the ideal valuation v of A has centre $\mathbf{p}$, then v is a Krull valuation of A if and only if it is a Krull valuation of $Q = A_{\mathbf{p}}$, and $\dim_A v = \dim_Q v$. Hence we can assume A is local and that $\mathbf{p}$ is its maximal ideal. Let B be the integral closure of Q in T(Q), that is, the set of x in F such that $Q \supseteq x\mathbf{p}^r$ for some r. Then, by Theorem 7.14, v is an ideal valuation of $L = B_{\mathfrak{n}}$, where $\mathfrak{n}$ is the centre of v on B, and, further, $\dim_L v = 0$. Also L is noetherian by Lemma 3.14 and the same lemma shows that L is either a discrete valuation ring or has grade >1. The latter is impossible, since, by i) it implies that $\dim_L v \geq 1$. If the former holds, then v is a Krull valuation of A by the definition at the beginning of section 2 of chapter 3.

To prove the last statement, we note that, if v is a Krull valuation of A, and $x \neq 0$ in A satisfies $v(x) > 0$, then v is associated with the filtration f_{xA} on A,and so is an ideal valuation of A. Further, it satisfies $\dim_A v = 0$ by Theorem 3.21.

COROLLARY. *Let* v *be an ideal valuation on* A, *centre* $\mathbf{p}$. *Let* $Q = A_{\mathbf{p}}$ *have residue field* k. *Then if* $x_1,\dots,x_r$ *are elements of* O_v *whose images in the residue field* K_v *of* v *form a transcendence basis of* K_v *over* k, *then* v *is a Krull valuation of* $B = A[x_1,\dots,x_r]$.

It is clear that, if $\mathfrak{p}$ is the centre of v on B, then K_v is algebraic over $B/\mathfrak{p}B_{\mathfrak{p}}$. Hence $\dim_B v = 0$, and the statement follows from the theorem.

2. Ideal valuations of finitely generated extensions.

In this section we are concerned with the case where A is a noetherian domain, with field of fractions F, and B is a finitely generated extension of A (abbreviated below to f.g.e.) whose field of fractions E need not equal F. It will be convenient in this section to consider the trivial valuation on a field E or F to be an ideal valuation of any noether domain with that field as its field of fractions. This is somewhat at variance with our earlier use of the term ideal valuation, where this case was implicitly excluded, but in the present section the extension of the use of the term will simplify the statements of certain theorems.

Below, if v is a valuation, K_v will denote the residue field of v.

We now introduce some terminology directed at shortening the statement of the conditions of the following theorems. This requires some preliminary explanation. Suppose that E is a finitely generated field extension of a field F of transcendence degree r, and that w is a discrete rank 1 valuation on E whose restriction to F is v, which may be trivial. Next, suppose that we choose elements $y_1',\dots,y_s'$ of K_w which are algebraically independent over K_v and $y_1,\dots,y_s$ are elements of O_w with these elements as images in K_w. Then $y_1,\dots y_s$ are algebraically independent over F. For, suppose that $f(Y_1,\dots,Y_s)$ is a non-zero polynomial, coefficients in F, such that $f(y_1,\dots,y_s) = 0$. If v is trivial, then $f(y_1',\dots,y_s') = 0$ in K_w which is a contradiction. If v is not trivial, by multiplying by a suitable power of the generator of the maximal ideal of O_v, we can assume that the coefficients of f are all in O_v and that at least one is not in the maximal ideal of O_v. Reducing modulo the maximal ideal of O_v then gives us an algebraic relation over K_v between $y_1',\dots,y_s'$, contrary to their choice. Hence $s \leq r$. Now suppose v is trivial and w is not. Then any element y of E such that $w(y) < 0$ must be transcendental over $F(y_1,\dots,y_s)$, since the construction above implies that $F(y_1,\dots,y_s)$ is contained in O_w and hence, if y is algebraic over $F(y_1,\dots,y_s)$, it is integral over O_w, implying that $w(y) \geq 0$. Hence in this case $s \leq r-1$. We now give a definition.

DEFINITION. *Let* E *be a finitely generated extension of a field* F *such that* $\text{trans.deg}_F E = r$. *Let* w *be a discrete rank* 1 *valuation on* E *or the trivial valuation. Let* v *be the restriction of* w *to* F *(which may also be trivial). Then we say that* w *is maximal over* F *if either*

a) K_w *has transcendence degree* r *over* K_v, *or,*

b) v *is trivial, but* w *is not, and* K_w *has transcendence degree* $r-1$ *over* F.

Note that the definition requires that the transcendence degree of K_w over K_v should be as large as possible.

THEOREM 7.21. *Let* A *be a noetherian domain, field of fractions* F, *and let* $B \supseteq A$ *be a noetherian domain whose field of fractions* E *is finitely generated over* F. *Let* $r = \text{trans.deg}_F E$ *and let* w *be an integer-valued valuation on* E, ≥ 0 *on* B. *Then, if* w *is maximal over* F, w *is an ideal valuation of* B.

If the restriction v of w to F is trivial, adjoin to F elements $y_1,\dots,y_{r-1}$ of O_w whose images in K_w form a transcendence basis of that field over F. Then $y_1,\dots,y_{r-1}$ are algebraically independent over F. Let $y_r,\dots,y_m$ be further elements of O_w such that $E = F(y_1,\dots,y_m)$. Then, if $\mathfrak{P}$ is the centre of w on $C' = F[x_1,\dots,x_m]$, and K is the field of fractions of $C'/\mathfrak{P}$, K_w is algebraic over K. Now apply the dimension inequality to F, C', $\mathfrak{P}$, to obtain

$$0 + r \geq \text{ht}\mathfrak{P} + \text{trans.deg}_F K.$$

But, as K_w is algebraic over K, $\text{trans.deg}_F K = r-1$. Hence $\mathfrak{P}$ has height 1 and w is an ideal valuation of C'. But C' is the ring of fractions $A[y_1,\dots,y_m]_S$ where S is the set of non-zero elements of A. Hence w is an ideal valuation of $C = A[y_1,\dots,y_m]$, which has field of fractions E. It now follows by Theorem 7.14 i), that it is an ideal valuation of $B[y_1,\dots,y_m]$ which also has field of fractions E. Finally, by 7.14 ii), w is an ideal valuation of B.

In case ii) we can, by adjoining a suitable set of elements $x_1,\dots,x_n$ of O_v to A, assume that v has centre a height 1 prime ideal of A (and, is a Krull valuation of the extended A). This replaces B by $B[x_1,\dots,x_n]$, but 7.14 i) and ii) show that w is an ideal valuation of B if and only if it is an ideal valuation of $B[x_1,\dots,x_m]$. We also note that since v is a Krull valuation of A, K_v is a finite algebraic extension of the field of fractions k of $A/\mathfrak{p}$, where $\mathfrak{p}$ is the centre of v on A. We now choose $y_1,\dots,y_r$ in O_w whose images in K_w form a transcendence basis of K_w over K_v (and hence over k). Then $y_1,\dots,y_r$ are algebraically independent over F. We now choose $y_{r+1},\dots,y_m$ in O_w, so that $E = F(y_1,\dots,y_m)$. Let $C = A[y_1,\dots,y_m]$ and let $\mathfrak{P}$ be the centre of w on C and K be the field of fractions of C. Then K_w is a finite algebraic extension of K, and applying the altitude inequality to A,C $\mathfrak{P}$, we obtain

$$ht\mathfrak{p} + r \geq ht\mathfrak{P} + \text{trans.deg}_k K.$$

But as K_w has transcendence degree r over K_v and K_w,K_v are respectively algebraic extensions of K,k, it follows that $\text{trans.deg}_k K = r$, and hence $ht\mathfrak{P} = 1$. Hence w is an ideal valuation of C by 7.11, hence of $B[y_1,\dots,y_m]$ by 7.14 i) and finally of B by 7.14 ii).

THEOREM 7.22. *Let* A *be a noetherian domain, field of fractions* F *and let* E *be a finitely generated field extension of* F. *Let* B *be a finitely generated ring extension of* A *whose field of fractions is* E. *Then the ideal valuations of* B *are precisely those extensions of ideal valuations of* A *to* E *which are* ≥0 *on* B *and which are maximal over* F.

Theorem 7.21 shows that an extension of an ideal valuation v of A which is ≥0 on B and which is maximal over F is an ideal valuation of B whether B is finitely generated over A or not. Hence we only have to prove the converse, i.e., we are given that w is an ideal valuation of B and we have to prove that it is an ideal valuation of any sub-ring A of B over which B is finitely generated, and, further, is maximal over the field of fractions of A.

We commence with a simple observation. Suppose that we have fields $E \supseteq E' \supseteq F$, and a discrete integer-valued valuation w on E. Then w is maximal over F if and only if it is maximal over E' and its restriction to E' is maximal over F. The proof is left to the reader.

We now suppose that $B = A[x_1,...,x_n]$. We will use induction on n to reduce the proof to the case n = 1, the case n = 0 being trivial. We write E_i for the field of fractions of $B_i = A[x_1,...,x_{n-i}]$ and w_i for the restriction of w to E_i. We now show that the inductive step from i to i+1 follows from the case n = 1. Suppose we have proved that w is maximal over E_i and its restriction to E_i is an ideal valuation of B_i. Then the case n = 1 will imply that w_i is maximal over E_{i+1} and that the restriction of an ideal valuation of B_i to E_{i+1} is an ideal valuation of B_{i+1}. Putting these together shows that w is maximal over E_{i+1} and is an ideal valuation of B_{i+1}. Hence we now need only consider the case n = 1, and write x for x_1. There are two possibilities.

i) x is algebraic over F. Then, for a suitable $a \neq 0$ in A, z = ax is integral over A and B = A[z][z/a]. Hence by Theorem 7.14, w is an ideal valuation of A[z]. Hence it is sufficient to prove the result with the added assumption that x is integral over A.

Write v for the restriction of w to v. Since w is non-trivial and E = F(x) is algebraic over F, v is non-trivial. Let $\mathfrak{P}$, $\mathfrak{p}$ be the centres of w, v on B, A respectively, so that $\mathfrak{p} = \mathfrak{P} \cap A$. Let K, k be the fields of fractions of $B/\mathfrak{P}$, $A/\mathfrak{p}$ respectively. Then, as $B/\mathfrak{P}$ is a finite module over $A/\mathfrak{p}$, K is a finite extension of k. Since E is a finite extension of F, it follows from Lemma 7.14, applied to $Q = O_v$, that K_w is a finite algebraic extension of K_v (and hence w is maximal over F). Further, as w is an ideal valuation of B, K_w is a finitely generated extension of K and hence K_v is a finitely generated extension of k. Now choose $z_1,...,z_r$ in O_v whose images in K_v form a transcendence base of K_v over k. Then their images in K_w form a transcendence base of K_w over K. Hence by the Corollary to Theorem 7.18, w is a Krull valuation of $B[z_1,...,z_r]$. But $B[z_1,...,z_r]$ is an integral extension of $A[z_1,...,z_r]$, and hence, by Theorem 7.17, its restriction v to F is a Krull valuation of $A[z_1,...,z_r]$ and therefore, by Theorem 7.11, an ideal valuation of A.

ii) x is transcendental over F. First suppose that w is trivial on F. Then w is an ideal valuation of F[x]. But F[x] is a principal ideal domain, and if $\mathfrak{P}$ is any non-zero prime ideal of $F[x]$, (in particular if $\mathfrak{P}$ is the centre of w on $F[x]$), $F[x]_{\mathfrak{P}}$ is a discrete valuation ring, whose residue field is a finite algebraic extension of F. If P is the centre of w, this shows that w is maximal over F.

We now assume that the restriction of w to F is a non-trivial valuation v. Then $w(z) \geq 0$ on $O_v[x]$, and so, by Theorem 7.14(ii), w is an ideal valuation of $O_v[x]$. Further, if p denotes the generator of the maximal ideal of O_v, $w(p) > 0$. Hence the centre $\mathfrak{P}$ of w on $O_v[x]$ contains p. We now distinguish two possibilities.

a) $\mathfrak{P} = pO_v[x]$. Then $O_v[x]/\mathfrak{P} \cong K[X]$, where X is an indeterminate over K. Let

$$y = a_0 + a_1x + ...+ a_rx^r,$$

with a_i in F. Let $m = \operatorname{Min}(v(a_i)/v(p))$, which is an integer, and write $b_i = p^{-m}a_i$ and $z = p^{-m}y$. Then z belongs to $O_v[x]$ but not to $\mathfrak{P}$, implying that $w(z) = 0$ and hence $w(y) = mv(p)$. Hence, if w is normalised, we must have $v(p) = 1$. Now suppose y above belongs to $B = A[x]$, so that a_i belongs to A for each i. Then the above shows that w is the unique graded extension of v to B defined by $w(x) = 0$. Since $w(x) = 0$, $w(z) \geq 0$ on $A[x,x^{-1}]$ and is an ideal valuation of this ring. Suppose that w is associated with an ideal J of $A[x,x^{-1}]$ (i.e., is associated with the filtration f_J). As w is graded, it follows that, if J_h is the smallest graded ideal of $A[x,x^{-1}]$ containing J, then $w(J_h) = w(J)$, and, by Theorem 6.15, w is associated with J_h. Hence we can assume that J is graded. But, as J is graded,

$$J = J'a[x,x^{-1}], \quad \text{where } J' = A\cap J.$$

But this, expressing y as above, implies that

$$f_J(y) = \operatorname{Min}(f_{J'}(a_i)),$$

and hence that, if the valuations associated with J' are $v_1,...,v_s$, and $e_j = v_j(I')$, then

$$\mathbf{f}_J(y) = \operatorname{Min}(v_j(a_i)/e_j),$$

the minimum being taken over $i = 0,...,r$ and $j = 1,...,s$. If we now define w_j to be the graded extension of v_j to B defined by $w_j(x) = 0$, then this can be written as

$$\mathbf{f}_J(x) = \operatorname{Min}(w_j(y)/e_j)$$

and hence w(z) as a valuation associated with J is equal to $w_j(z)$ for some j (since both are assumed normalised). Hence $v = v_j$ and so is associated with J' and hence is an ideal valuation of A. Further, we have proved that K_w is a pure transcendental extension of $K = K_v$, and hence w is maximal over F.

b) $\mathfrak{P} \neq pO_v[x]$. Then, as $O_v[x]/pO_v[x] \cong K[X]$, where $K = K_v$, a prime ideal $\mathfrak{P}$ properly containing $pO_v[x]$ is maximal, $\mathfrak{P}/pO_v[x]$ is principal, and $O_v[x]/\mathfrak{P}$ is a finite algebraic extension of K. Hence if $Q = O_v[x]_{\mathfrak{P}}$, Q is a two-dimensional regular local ring (and in particular is C-M). Since w is an ideal valuation of Q, Theorem 6.22(ii) implies that K_w has transcendence degree 1 over the residue field of Q and hence over K, i.e., w is maximal over F. Now let z be an element of O_w whose image in K_w is transcendental over K. Then z is transcendental over F, and hence $E = F(x)$ is algebraic over F(z). Then applying i) above to A[z], A[x], the restriction v' of w to F(z) is an ideal valuation of A[z]. We now consider the rings A, A[z] and the valuations v,v'. Since the image of z in $K_{v'}$ is transcendental over K, it follows that the centre of v' on $O_v[x]$ is $pO_v[x]$, and hence we are in the situation of a) above. Hence v, as the restriction of v' to A, is an ideal valuation of A.

3. Applications.

In this section we will be concerned with quasi-heights and their relationship with both ideal valuations and chains of prime ideals. The results we present are not new, being translations into a different language of results of Nagata, Ratliff and others. Our principal tools are the results of the last two sections together with Theorem 6.24. Since the latter is constantly referred to, we, repeat the statement in a shortened form for the reader's convenience.

If $\mathfrak{p}$ *is a prime ideal of a noetherian ring* A, *then* h *is a quasi-height of* $\mathfrak{p}$ *if and only if there exists an ideal valuation* v *of* A *with centre* $\mathfrak{p}$, *whose dimension over* A *is* h - 1.

Note that the above is stated for a noetherian ring, whereas throughout this chapter we have restricted attention to noetherian domains. We will, in fact, deal for the most part with domains, but the extension of the results for this case to the general case is fairly easy if we note that h is a quasi-height of $\mathfrak{p}$ if and only if it is the quasi-height of the ideal $\mathfrak{p}/\mathfrak{P}$ of the domain $A/\mathfrak{P}$ for some minimal prime ideal $\mathfrak{P}$ of A contained in $\mathfrak{p}$.

We commence with a result which contains the altitude inequality. However, this result has been used in setting up the background of the proof, and so this is not an independent proof.

THEOREM 7.31. *Let* A,B $\supseteq$ A *be noetherian domains,* B *being finitely generated over* A, *and let* F, E *be the fields of fractions of* A,B. *Let* $n = \text{trans.deg}_F E$. *Let* $\mathfrak{P}$ *be a non-zero prime ideal of* B *and write* $\mathfrak{p}$ *for* $\mathfrak{P} \cap A$. *Let* K,k *denote the fields of fractions of* $B/\mathfrak{P}$ *and* $A/\mathfrak{p}$ *respectively. Finally let* $\text{trans.deg}_k K = n - t$. *Then*

i) *if* $\mathfrak{p} = (0)$, *then the only quasi-height of* $\mathfrak{P}$ *is* t,

ii) *if* $\mathfrak{p} \neq 0$, *then the quasi-heights of* $\mathfrak{P}$ *are contained in the set of numbers* h + t, *where* h *ranges over the quasi-heights of* $\mathfrak{p}$.

Let h' be a quasi-height of $\mathfrak{P}$. Then by Theorem 6.24, there exists an ideal valuation w on B such that $\dim_B w = h' - 1$. First suppose that $\mathfrak{p} = (0)$. Then k = F and, by Theorem 7.24, K_w has transcendence degree n - 1 over F. Hence $h'-1 = \text{trans.deg}_K K_w = \text{trans.deg}_F K_w - \text{trans.deg}_F K = (n-1) - (n-t) = t-1$, whence h' = t.

Now suppose that $\mathfrak{p} \neq (0)$ and let v be the restriction of w to F so that v is an ideal valuation of A. Hence, if $h = \dim_A v + 1 = \text{trans.deg}_k K_v + 1$, h is a quasi-height of $\mathfrak{p}$. We also note that K_w has transcendence degree n over K_v by Theorem 7.24. Hence

$$h' = \text{trans.deg}_K K_w + 1 = \text{trans.deg}_k K_w + 1 - (n-t) = h + n - (n-t) = h + t$$

yielding the required result.

(Note that the altitude inequality follows by taking $h' = \text{ht}\mathfrak{P}$ and noting that every quasi-height of $\mathfrak{p}$ is $\leq \text{ht}\mathfrak{p}$.)

Our next lemma forms the basis of an inductive procedure for considering saturated chains of prime ideals. It will be generalised in the last theorem of this section.

LEMMA 7.32. *Let* $(Q,\mathfrak{m},k,d)$ *be a local domain,* $\mathfrak{p}$ *be a height* 1 *prime ideal of* Q. *Then*

i) *if* $\dim(Q/\mathfrak{p}) = 1$, 2 *is a quasi-height of* $\mathfrak{m}$,

ii) *if* h *is a quasi-height of* $Q/\mathfrak{p}$, h+1 *is a quasi-height of* $\mathfrak{m}$.

i) Write Q' for $Q/\mathfrak{p}$. Choose $x \neq 0$ in $\mathfrak{p}$ and y in $\mathfrak{m}$ not in any prime ideal associated with the radical of xQ. Let I_n denote the integral closure of the ideal $xQ+y^nQ$. Finally let $I_n' = I_n + \mathbf{p}/\mathbf{p}$, and y' denote the image of y in Q'. Then I_n' is contained in the ideal $(y'^nQ')^*$. But, by Corollary i) to Theorem 2.33, the intersection of the ideals $(y'^nQ')^*$ is zero. Hence, if z is an element of Q not in $\mathbf{p}$, z does not belong to I_n for n sufficiently large.

Now suppose that $I_n : \mathbf{m} = I_n$. Then, as $\mathbf{p} + I_n$ properly contains $\mathbf{p}$, it is $\mathbf{m}$-primary, and hence $I_n : \mathbf{p} = I_n : (\mathbf{p}+I_n) = I_n$. This implies that $I_n : \mathbf{p}^m = I_n$ for all m. But, $I_n : \mathbf{p}^m \supseteq xQ : \mathbf{p}^m$ and the latter contains an element z, independent of n, not in $\mathbf{p}$ if m is sufficiently large. It follows that $I_n : \mathbf{m} \neq I_n$ if n is large enough for I_n not to contain z and hence $\mathbf{m}$ is associated with I_n if n is sufficiently large. But an ideal associated with the integral closure of an ideal is the centre of a valuation v associated with that ideal. Further, if $J = xQ + y^nQ$, then $G(J)/\mathbf{m}G(J) + uG(J)$ is isomorphic to $k[X_1,X_2]$ and $\mathbf{m}G(J) + uG(J)$ is the centre of a Krull valuation of $G(J)$ extending v. Hence as the residue field K_V of V is an algebraic extension of the field of fractions of $G(J)/\mathbf{m}G(J)+uG(J)$, it has transcendence degree 2 over k. But the field of fractions of $G(J)$ has transcendence degree 1 over that of Q, and hence by 7.23, K_V has transcendence degree 1 over K_v, whence the latter has transcendence degree 1 over k. Since v is an ideal valuation of Q, 6.24 implies that 2 is a quasi-height of Q.

ii) Let F, F' be the fields of fractions of Q, Q' respectively. By the Corollary to 7.11, we can choose $x_1,\dots,x_r$ in $Q_{\mathbf{p}}$ with the following property. If x_i' denotes the image of x_i in F', and if B' denotes $Q'[x_1',\dots,x_r']$, then there is a height 1 prime ideal $\mathfrak{P}'$ of B' such that

a) $\mathfrak{P}'\cap Q' = \mathbf{m}'$, where $\mathbf{m}' = \mathbf{m}/\mathbf{p}$,

b) the field of fractions of $B'/\mathfrak{P}'$ has transcendence degree $h-1$ over k.

Let $B = Q[x_1,\dots,x_r]$ and let $\mathfrak{P}$ be the inverse image of $\mathfrak{P}'$ under the map of B onto B' induced by the map of $Q_{\mathbf{p}}$ onto F'. Let $\mathfrak{P}'' = B\cap\mathbf{p}Q_{\mathbf{p}}$, so that $B_{\mathfrak{P}''} = Q_{\mathbf{p}}$ and $\mathfrak{P}''\cap Q = \mathbf{p}$. Localise B at $\mathfrak{P}$ and let the resulting ring be denoted by Q^*, its maximal ideal by $\mathbf{m}^*$ and its residue field by k^*. Then we have the following:

α) k^* is isomorphic to the field of fractions of $B'/\mathfrak{P}'$, and so has transcendence degree $h - 1$ over k;

β) if $\mathfrak{p}^* = \mathfrak{p}Q_{\mathfrak{p}} \cap Q^*$, then $Q^*/\mathfrak{p}^*$ is isomorphic to $B'_{\mathfrak{p}}$, and so has dimension 1;

γ) $\mathfrak{p}^*$ has height 1, since $Q^*_{\mathfrak{p}^*} = Q_{\mathfrak{p}}$.

By part i) it now follows that 2 is a quasi-height of Q^*. Applying 7.31 to Q and B, it follows that $2 + h - 1 = h + 1$ is a quasi-height of Q.

COROLLARY. *Let* $\mathfrak{m} = \mathfrak{p}_1 \supset \ldots \supset \mathfrak{p}_h \supset (0)$ *be a saturated chain of prime ideals of a local domain* Q *with maximal ideal* $\mathfrak{m}$. *Then* h *is a quasi-height of* $\mathfrak{m}$.

We prove this by induction on h, the case h = 1 being immediate. The result then follows by applying the inductive hypothesis to $Q/\mathfrak{p}_h$.

For our next theorem we need to introduce some notation. Let $(Q,\mathfrak{m},k,d)$ be a local ring, $X_1,\ldots,X_n$ be a set of indeterminates over Q. Then we denote by Q_n and $Q_{(n)}$ the local rings obtained by localising $S = Q[X_1,\ldots,X_n]$ at $\mathfrak{m}S$ and $(\mathfrak{m},X_1,\ldots,X_n)S$ respectively. The maximal ideals of Q_n and $Q_{(n)}$ are $\mathfrak{m}_n$ and $\mathfrak{m}_{(n)}$. Further, the minimal prime ideals of Q_n and $Q_{(n)}$ are the ideals $\mathfrak{P}Q_n$ and $\mathfrak{P}Q_{(n)}$, where P ranges over the minimal prime ideals of Q. Since a quasi-height of $\mathfrak{m}$ is a quasi-height of $\mathfrak{m}/\mathfrak{P}$ for a suitable minimal prime ideal $\mathfrak{P}$ of Q, and conversely, it follows that in discussing quasi-heights it is sufficient to consider the case where Q is a domain.

THEOREM 7.33. i) *The quasi-heights of* $\mathfrak{m}$, $\mathfrak{m}_n$ *are the same, while*

ii) h *is a quasi-height of* $\mathfrak{m}$, $n + h$ *is a quasi-height of* $\mathfrak{m}_{(n)}$ *and these are all the quasi-heights of* $Q_{(n)}$. *Further,*

iii) *if* n *is sufficiently large, there exists a saturated chain of prime ideals*

$$\mathfrak{m}_{(n)} = \mathfrak{p}_1 \supset \ldots \supset \mathfrak{p}_{n+h} \supset \mathfrak{P}_{(n)},$$

of length $n + h$ *in* $Q_{(n)}$, $\mathfrak{P}$ *being a minimal prime ideal of* Q.

As stated above, we need only consider the case where Q is a domain, and we will make this restriction throughout this proof. Note that in iii) this implies that $\mathfrak{P} = (0)$.

i). Since $Q_n/\mathfrak{m}_n = k(X_1,\ldots,X_n)$, every quasi-height of $\mathfrak{m}_n$ is a quasi-height of $\mathfrak{m}$ by 7.31. Conversely, using 6.24, there is an ideal valuation v of Q centre $\mathfrak{m}$ such that $\dim_Q v = h - 1$. Let V be the graded extension of v to $Q[X_1,\ldots,X_n]$ defined by $V(f(X_1,\ldots,X_n)) = \mathrm{Min}\, v(a)$ where a runs over the coefficients of the polynomial f. Then V is an ideal valuation of $\mathfrak{m}_n$ by 7.23, and its residue field has dimension h-1 over $Q_n/\mathfrak{m}_n$. Hence h is a quasi-height of $\mathfrak{m}_n$.

ii). A straightforward induction reduces the result to the case n = 1. Suppose that h is a quasi-height of $\mathfrak{m}$. Then, as $Q = Q_{(1)}/X_1Q_{(1)}$, it follows from Lemma 7.32 that h + 1 is a quasi-height of $Q_{(1)}$. Next suppose that h' is a quasi-height of $Q_{(1)}$. Then, as $Q_{(1)}/\mathfrak{m}_{(1)} = k$ and the field of fractions of $Q_{(1)}$ has transcendence degree 1 over that of Q, it follows that h' - 1 is a quasi-height of Q. This completes the proof.

iii). Let h be a quasi-height of Q. Then, by the Corollary to Theorem 7.11, there exists a finitely generated extension $B = Q[x_1/x_r,\ldots,x_{r-1}/x_r]$, where $x_1,\ldots,x_r$ are elements of $\mathfrak{m}$, and a height 1 prime ideal $\mathfrak{p}$ of B meeting Q in $\mathfrak{m}$, such that the field of fractions K of $B/\mathfrak{p}$ has transcendence degree h - 1 over k. Now let t be an indeterminate over Q, write u for t^{-1}, and let $G = Q[tx_1,\ldots,tx_r,u]$ (that is, G = G(J), where J is the ideal generated by $x_1,\ldots,x_r$). Note that the ring of fractions $G[u^{-1}]$ of G is simply B[t,u]. Let $\mathfrak{P}$ be the graded prime ideal $\mathfrak{p}B[t,u]\cap G$ of G. Then the field of fractions of $G/\mathfrak{P}$ is K(t) and so has transcendence degree h over k. Further, $\mathfrak{P}$ has height 1. We now localise G at its maximal graded ideal $\mathfrak{m}G + uG + (x_1t,\ldots,x_rt)G$, and denote the resulting ring by Q' and the ideal $\mathfrak{P}Q'$ by $\mathfrak{P}'$ so that $\mathfrak{P}'$ has height 1. Now consider $Q'/\mathfrak{P}'$. This is a localisation of a finitely generated extension of k of transcendence degree h, and its residue field is isomorphic to k. Hence by 7.31 i), it follows that the only quasi-height of $Q'/\mathfrak{P}'$ is h, and hence any saturated chain of prime ideals joining the zero ideal of $Q/\mathfrak{P}'$ to its maximal ideal has length h. We can extend this by 1 and obtain a saturated chain of prime ideals joining the zero ideal of Q' to its maximal ideal of length h + 1.

Now we consider the map of $Q[X_1,\ldots,X_{r+1}]$ onto G in which $X_i \to x_it$ if $i\leq r$, and $X_{r+1} \to u$. This map clearly extends to a map of $Q_{(r+1)} \to Q'$. Let $\mathfrak{P}''$ be the kernel.

Then by the above, we have a saturated chain of prime ideals of length h + 1 joining $\mathfrak{P}''$ to $\mathfrak{m}_{(r+1)}$. Now localise $Q_{(r+1)}$ at $\mathfrak{P}''$ obtaining a ring Q". Since Q is a sub-ring of Q', $\mathfrak{P}''$ meets Q in (0). Hence Q" can be considered as a localisation of $F[X_1,...,X_{r+1}]$ at a prime ideal, where F is the field of fractions of Q. Now the residue field of Q" is the field of fractions of G,which is isomorphic to F(t), and so has transcendence degree 1 over F. Applying 7.31 i) again, we see that the only quasi-height of m" is r and hence every saturated chain of prime ideals joining (0) to $\mathfrak{P}''$ has length r. Putting the two saturated chains we have obtained together, we obtain one of length h + r + 1 joining (0) and hence our result is proved with n = r + 1.

The final theorem of this section includes Lemma 7.32 (ii) as a special case.

THEOREM 7.34. *Let* $(Q,\mathfrak{m},k,d)$ *be a local ring,* $\mathfrak{p}$ *be a prime ideal of* Q, h *be a quasi-height of* $\mathfrak{m}/\mathfrak{p}$ *and* h' *be a quasi-height of* $\mathfrak{p}$. *Then* h + h' *is a quasi-height of* $\mathfrak{m}$.

Since h' is a quasi-height of $\mathfrak{p}$ if and only if it is a quasi-height of $\mathfrak{p}/\mathfrak{P}$ for some minimal prime ideal $\mathfrak{P}$ of Q, we may assume that Q is a domain.

Let $Q' = Q/\mathfrak{p}$, $Q'' = Q_{\mathfrak{p}}$. Then, applying Theorem 7.33 to Q", there is a saturated chain of prime ideals of length h' + m joining (0) to the prime ideal $\mathfrak{p}_{(m)} = (\mathfrak{p},X_1,...,X_m)$ of $Q_{(m)}$ for any sufficiently large m and $Q_{(m)}/\mathfrak{p}_{(m)} = Q/\mathfrak{p}$. The chain remains stable if we adjoin further indeterminates. But for n sufficiently large we have a saturated chain of length h + n joining $\mathfrak{p}_{(m)}Q_{(m+n)}$ to $\mathfrak{m}_{(m+n)}$. Hence we have a saturated chain of length m + n + h + h' joining the zero ideal of $Q_{(m+n)}$ to $\mathfrak{m}_{(m+n)}$. This, by 7.33 ii) and the Corollary to 7.33, implies that h + h' is a quasi-height of Q.

4. More on the rings Q_n.

In this section we consider further properties of the local rings Q_n, since we will make extensive use of these rings later.

We commence by recalling the notions of flatness and faithful flatness. We suppose that A is a noetherian ring and that M is an A-module, which is not assumed to be finitely generated. Then M is said to be a *flat* A-module if, given any exact sequence of A-modules

$$0 \longrightarrow N' \longrightarrow N \longrightarrow N'' \longrightarrow 0$$

the sequence

$$0 \longrightarrow M\otimes N' \longrightarrow M\otimes N \longrightarrow M\otimes N'' \longrightarrow 0$$

is also exact, where we have written $M\otimes N$ for $M\otimes_A N$. A ring extension B of A is said to be a *flat* extension if B is a flat A-module. Examples of flat extensions of A are rings of fractions A_S and rings of polynomials $A[X_i|i \in I]$, where X_i is an indeterminate and I is a set, finite or infinite. We also note that if B is a flat extension of A and C is a flat extension of B then C is a flat extension of A. As a result, we see that Q_n is a flat extension of Q. Next we recall that an extension B of A is a faithfully flat extension of A if it is a flat extension and if $B\otimes M = 0$ implies that $M = 0$, where M is any A-module. If A is a local ring $(Q, \mathfrak{m}, k, d)$, then it is both necessary and sufficient that $B \neq \mathfrak{m}B$ for this to be the case. Since $Q_n/\mathfrak{m}Q_n$ is isomorphic to the field $k(X_1,\ldots,X_n)$ and so is certainly non-zero, Q_n is a faithfully flat extension of Q. In fact, much more is true. We have seen that $Q_n/\mathfrak{m}Q_n$ is a field.

(7.41) *Then, if* M *is a* Q*-module of finite length, and we denote by* M_n *the* Q_n*-module* $Q_n\otimes M$, *then the length of the* Q_n*-module* M_n *is equal to the length of the* Q*-module* M.

We now briefly recall some consequences of flatness and faithful flatness of an extension.

(7.42) *Suppose that* A *is a noetherian ring, that* B *is a flat extension of* A, *and that* I,J *are ideals of* A. *Then we have the following:*

i) $(I\cap J)B = IB\cap JB$; ii) $(I:J)B = IB:JB$.

Our next group of results goes beyond those which are consequences of Q_n being a faithfully flat Q-module.

We list them together as follows.

(7.43) *Let* $(Q,\mathfrak{m},k,d)$ *be a local ring and let* J *be an ideal of* Q. *Then:*

i) *the rings* $(Q/J)_n$ *and* Q_n/JQ_n *are isomorphic,*

ii) *if* J *is prime so is* JQ_n,

iii) *if* M *is a finitely generated* Q*-module,* M_n *denotes* Q_n*-module* $Q_n \otimes_Q M$*, and* $(0) = N_1 \cap \ldots \cap N_s$ *is an irredundant primary decomposition of the zero sub-module* (0) *of* M*, then*

$$(0) = N_{1n} \cap \ldots \cap N_{sn}$$

is an irredundant primary decomposition of the zero sub-module of M_n*. Further, if* N_i *has radical* $\mathbf{p}_i$*,* N_{in} *has radical* $\mathbf{p}_i Q_n$*.*

Result iii) is a consequence of ii) or the fact that Q_n is a flat extension of Q. For a proof of this, the reader is referred to [BAC] chapter IV, proposition 11 on p.157. More generally, as general references for flatness we refer the reader to [BAC] chapter 1 and [M], chapters 2 and 8.

For our final group of results we first require a lemma.

LEMMA 7.44. *Let* $(Q, \mathfrak{m}, k, d)$ *be a local ring, and let* f *be a noether filtration on* Q*, and let* f_n *be the noether filtration on* Q_n *defined by the sequence of ideals* $I_m(f)Q_n$*. Then, if* $v_1, \ldots, v_s$ *are the valuations associated with* f*, those associated with* f_n *are* $V_1, \ldots, V_s$ *which are defined on* $Q[X_1, \ldots, X_n]$ *by the rule*

$$V_i(p(X_1, \ldots, X_n)) = \mathrm{Min}\,(v_i(a) \mid a \text{ is a coefficient of } p)$$

Further, $(f_n)^* = (f^*)_n$*.*

Suppose x is an element of Q_n satisfying

$$V_i(x) \geq m v_i(f),\ i = 1, \ldots, s.$$

Now we can write x in the form p/q, where p, q are polynomials over Q in $X_1, \ldots, X_n$, with at least one coefficient of q being a unit of Q. It then follows that $V_i(x) = V_i(p)$ and hence every coefficient of p satisfies

$$v_i(a) \geq m v_i(f),\ i = 1, \ldots, s,$$

and hence $x \in I_m(f^*)Q_n$. The converse being immediate, it follows that $I_m(f^*)Q_n$ is the set of elements of Q_n satisfying the inequalities

$$V_i(x) \geq m v_i(f),\ i = 1, \ldots, s,$$

and this implies that the valuations $V_{i(x)}$ are the valuations associated with the filtration determined by the sequence of ideals $I_m(f^*)Q_n$.

Now consider the filtration on Q_n defined by the sequence of ideals $I_m(f^*)Q_n$. This is $[\mathrm{Min}(V_i(x)/v_i(f))]$, and so is an integrally closed filtration. It is $\geq f_n$ on Q_n since the latter is defined by the ideals $I_m(f)Q_n$, and hence $\geq f_n^*$. But $f_n^*(x) \geq m$ if x belongs to $I_m(f)Q_n$. Hence f_n^* is defined by the sequence of ideals $I_m(f^*)Q_n$, i.e., $(f^*)_n=(f_n)^*$. Further, this implies that the valuations associated with f_n are $V_1,\ldots,V_s$.

THEOREM 7.45. *Let* $(Q,\mathfrak{m},k,d)$ *be a local ring. Then if* Q *has any of the following properties*

a) Q *is Cohen-Macaulay (C-M),*

b) Q *is normal,*

c) Q *is analytically unramified,*

d) Q *is quasi-unmixed,*

then so does Q_n.

If Q is C-M, then Q contains a Q-sequence $(x_1,\ldots,x_d)$. But then, by 7.42 ii)

$$m(x_1,\ldots,x_{i-1})Q_n : x_i = (x_1,\ldots,x_{i-1})Q_n$$

for $i = 1,\ldots,d$, i.e., $(x_1,\ldots,x_d)$ is also a Q_n-sequence, and Q_n is a C-M ring.

If Q is normal, so is $Q[X_1,\ldots,X_n]$ and hence any ring of fractions of the latter, in particular, Q_n.

If Q is analytically unramified, then, if f is any noether filtration with radical $\mathfrak{m}$, for example, $f = f_{\mathfrak{m}}$, there exists a constant K such that $K \geq f^*(x) - f(x) \geq 0$ for all x by Theorem 5.32. (or 5.34). If we take $f = f_{\mathfrak{m}}$, this implies that $m^r \supseteq (m^{r+k})^*$ for all r. But it follows from Lemma 7.44 that f_n^* is defined by the sequence of ideals $(\mathfrak{m}^r)^*Q_n$ and applying the above, we see that

$$K \geq f_n^*(x) - f_n(X) \geq 0$$

with the same K and all x in Q_n. Then 5.34 implies that Q_n is analytically unramified.

If Q is quasi-unmixed, then Q has only one quasi-height and hence by Theorem 7.33 (i), Q_n also has only one quasi-height and so is quasi-unmixed.

8. THE MULTIPLICITY FUNCTION ASSOCIATED WITH A FILTRATION

1. Filtrations on a module.

Let A be a commutative ring, f be a filtration on A and let M be an A-module. An f-filtration on M is a function h(m) defined for all m in M, taking as values real numbers together with ∞, and satisfying the conditions,

i) $h(m - m') \geq \mathrm{Min}(h(m), h(m'))$ for all m, m' in M,

ii) $h(am) \geq h(m) + f(a)$ for all a in A, m in M.

We will be concerned below with the case where f is a noether filtration on a noetherian ring A, which takes only non-negative integer values together with ∞, M is a finitely generated A-module, and also h is restricted to take only non-negative integer values together with ∞. These restrictions will be in force unless otherwise stated.

Under these conditions, we can associate with an f-filtration on M a graded G(f)-module G(h,M) consisting of all finite sums

$$\Sigma m_r t^r$$

with $h(m_r) \geq r$, the sum running over a range $-p$ to q. We define the action of G(f) on G(h,M) in the obvious way, that is, by the distributive laws and

$$(at^r)(mt^s) = amt^{r+s}.$$

We then say that h is a *good* f-filtration on M if G(h,M) is a finitely generated G(f)-module.

It is clear that the f-filtration h determines, and is determined by, the sequence of sub-modules $M_r(h)$ consisting of all elements m of M such that $h(m) \geq r$. The condition ii) of the definition of an f-filtration is then equivalent to the statement that

$$M_{n+r}(h) \supseteq I_n(f).M_r(h).$$

Further, if h is good, and G(h,M) is generated by the elements $m_1 t^{r(1)}, \ldots, m_s t^{r(s)}$, then

$$M_n = I_{n-r(1)}(f)m_1 + \ldots + I_{n-r(s)}(f)m_s.$$

2. The multiplicity function of $\mathfrak{m}$-primary filtrations.

Now we impose restrictions on A and f. We first suppose that A is a local ring

$(Q,\mathfrak{m},k,d)$. Next we suppose that rad$f = \mathfrak{m}$. Then by Lemma 6.13, it follows that f has spread d. Further, the ideals $I_n(f)$ are $\mathfrak{m}$-primary, and we will also term f an $\mathfrak{m}$-primary filtration. Now suppose that M is a finitely generated Q-module and that h is a good f-filtration on M. Then the quotient modules M/M_r all have finite length. We now define $P(h,M,z)$ to be the formal power series $\Sigma l(M/M_{r+1})z^r$, where $l(N)$ denotes the length of N.

Since f has spread d, Theorem 6.12 implies that there exists a basic reduction of f, that is, a noether filtration $b(x)$ such that $b(x) \leq f(x)$ for all x in Q, b is equivalent to f, and is generated by d elements $x_1,\ldots,x_d$ with assigned weights $w(1),\ldots,w(d)$. Let X denote the set of elements $x_1t^{w(1)},\ldots,x_dt^{w(d)}$ of $G(f)$. Then $XG(f)$ is an irrelevant ideal of $G(f)$, i.e., the homogeneous components $(G(f)/\mathbf{X}G(f))_n$ of $G(f)/XG(f)$ are of finite length for all n and are zero if n is large.

We now consider the Koszul Complexes $K(f) = K(G(f,M),X)$, and $K(h,M) = K(G(h,M),X)$ as defined in section 2 of chapter 1. We first make some minor changes in notation. The set of elements **X** consists of the elements $x_1t^{w(1)},\ldots,x_dt^{w(d)}$ which are homogeneous of degrees $w(1),\ldots,w(d)$. We will therefore write $w(i)$ in place of f_i. Next suppose that $S = \{s(1) < s(2) < \ldots < s(p)\}$ is an ordered sub-set of $1,2,\ldots,d$. Then we will write $w(S)$ for $w(s(1)) + \ldots + w(s(p))$. This implies that the generator $u(S)$ of the free $G(f)$-module $K_p(f)$ has degree $w(S)$. We will also write G for $G(f)$ and $G(M)$ for $G(h,M)$. Note that $K_p(h,M) = K_p(f)\otimes_G G(M)$.

Now the homology modules $H_i(K(h,M))$ are graded modules annihilated by **X**, and each has homogeneous components of finite length. Further, if n is large, the homogeneous component of degree n of $H_1(G(h,M),\mathbf{X})$ is zero. If L(.) denotes length, we write

$$\chi(h,M,X,n) = \Sigma(-1)^i L(H_i(G(h,M),X)_n)$$

the sum being from 0 to d, and

$$\chi(h,M,\mathbf{X},z) = \Sigma\chi(h,M,\mathbf{X},n+1)z^n,$$

the latter sum being from $-\infty$ to ∞. However, by the remarks above, the coefficient of z^n is zero if n is large and positive.

Now, as a Q-module, $G(h,M)$ is a sub-module of the module $M[t,u]$ consisting of all finite sums $\Sigma m_r t^r$, $m_r \in M$, which can also be considered as $M \otimes Q[t,u]$, and, as such is also a $Q[t,u]$-module. We now consider the Koszul Complex $K(M[t,u],\mathbf{X})$. This is the same whether we consider $M[t,u]$ as a $G(f)$-module or as a $Q[t,u]$-module. It follows that the homology modules $H_p(M[t,u],\mathbf{X}) = H_p(K(M[t,u],\mathbf{X})$ are graded $Q[t,u]$-modules considered as $G(f)$-modules by restriction. We are now going to show that

$$H_p(M[t,u],\mathbf{X}) = H_p(M[t,u],\mathbf{x}) \qquad (*)$$

where $\mathbf{x} = (x_1,\dots,x_d)$, and so is a set of elements of degree zero of $Q[t,u]$. To distinguish the latter case from the former, we will denote the corresponding Koszul Complexes by $K^*(\mathbf{x})$ and $K^*(M[t,u],\mathbf{x})$. Thus $K_p{}^*(\mathbf{x})$ is a free $Q[t,u]$-module with generators $u^*(S)$, all of degree 0, defined for each ordered sub-set S of $(1,\dots,d)$ containing p elements. The boundary operators d,d^* in the complexes $K(M[t,u],\mathbf{X})$ and $K^*(M[t,u],\mathbf{x})$ are, respectively, defined by

$$d(u(S)) = \Sigma(-1)^j x_{s(j)} t^{w(s(j))} u(S_j)$$

and

$$d^*(u^*(S)) = \Sigma(-1)^j x_{s(j)} u^*(S_j)$$

where S_j is the ordered sub-set $s(1),\dots,s(j-1),s(j+1),\dots,s(p)$ obtained by deleting $s(j)$ from S, and in each case the sum is from $j = 1$ to $j = p$. Now suppose we define, for each p, a map $\boldsymbol{\theta}_p$: $K(Q[t,u],X) \to K^*{}_p(Q[t,u],x)$ by defining

$$\boldsymbol{\theta}_p(u(S)) = t^{w(S)} u^*(S).$$

This clearly satisfies

$$d^* \boldsymbol{\theta}_p = \boldsymbol{\theta}_{p-1} d$$

and, since t is a unit of $Q[t,u]$, the set of maps $\{\boldsymbol{\theta}_p\}$ defines an isomorphism of complexes. Tensoring with $M[t,u]$ then gives an isomorphism of complexes between $K(M[t,u],\mathbf{X})$ and $K^*(M[t,u],\mathbf{x})$ and hence an isomorphism of homology. Note also that

$$H_i(K^*(M[t,u],\mathbf{x}) = H_i(M,\mathbf{x}) \otimes_Q Q[t,u],$$

where $H_i(M,\underline{\mathbf{x}})$ is the ith homology module of the complex $K(Q,\mathbf{x}) \otimes_Q M$, $K(Q,\mathbf{x})$ being the ordinary Koszul complex of Q with respect to $\mathbf{x}$. It follows from the isomorphism that

$$L(H_i(K(M[t,u],\mathbf{X})_{n+1}) = L(H_i(M,\mathbf{x})),$$

the modules $H_i(M,\mathbf{x})$ being annihilated by $\mathbf{x}$ and so having finite length.

We now recall that the multiplicity function $e(\mathbf{x},M)$ with respect to the set $\mathbf{x}$ can be defined as

$$e(\mathbf{x}, M) = \Sigma(-1)^i L(H_i(M,\mathbf{x})).$$

We are now in a position to prove the main result of this section. We follow an argument first used by Auslander and Buchsbaum in [1959].

THEOREM 8.1 i). $P(h,M,z) = [(1-z^{w(1)})\ldots(1-z^{w(d)})(1-z)]^{-1}[e(\mathbf{x},M) - (1-z)h(z)]$ *where* $h(z)$ *is a polynomial in* z.

ii). *If* $e(f,M)$ *is defined to be the limit of* $(1-z)^{d+1}P(h,M,z)$ *as* z *tends to* 1 *from below, then*

$$e(f,M) = e(\mathbf{x},M)/w(1)\ldots w(d)$$

and depends only on M *and the equivalence class of* f.

To prove i) we define $K'(M,\mathbf{X})$ to be the quotient complex

$$K(M[t,u],\mathbf{X})/K(G(h,M),\mathbf{X}),$$

both the numerator and the denominator being considered as graded Q-modules. We now consider the complex $K'(M,\mathbf{X})$ in detail, considering its component of degree n. The term $K'_j(M,\mathbf{X})_n$ can be described as follows. Let J range over all sub-sets of $1,\ldots,d$ containing j elements. Then $K'_j(M,\mathbf{X})_n$ is the direct sum of Q-modules isomorphic to $M/M_{n-w(J)}$, one for each J, and so is of finite length. Hence

$$\Sigma(-1)^j L(H_j(K'(M,\mathbf{X})_n)) = \Sigma(-1)^j L(K'_j(M,\mathbf{X})_n).$$

Now write $\chi'(h,M,\mathbf{X},n)$ for either side of this equation and $\chi'(h,M,\mathbf{X},z)$ for the formal power series $\Sigma\chi'(h,L,X,n+1)z^n$. Then, since

$$\Sigma L(K'_j(M,\mathbf{X})_{n+1}) = (\Sigma z^{w(J)})P(h,M,z),$$

where the sum in brackets is over all ordered sub-sets J of $1,2,\ldots,d$ containing j elements

$$\chi'(h,M,\mathbf{X},z) = (1-z^{w(1)})\ldots(1-z^{w(d)})P(h,M,z).$$

Applying the additivity of the Euler-Poincaré characteristic to the exact sequence

$$0 \longrightarrow K(G(h,M),\mathbf{X}) \longrightarrow K(M[t,u],\mathbf{X}] \longrightarrow K'(M,\mathbf{X}) \longrightarrow 0,$$

we obtain

$$\chi'(h,M,\mathbf{X},z) = \chi(K(m[t,u],\mathbf{X}) - \chi(h,M,\mathbf{X},z) = e(M,x)\Sigma z^n - \chi(h,M,\mathbf{X},z)$$

the sum Σz^n being from $-\infty$ to $+\infty$, and the coefficients of z^n for n large and positive in the second term on the right-hand side being zero. Since the coefficients of negative powers on the left-hand side are all zero, we can replace the right-hand side by the sum of its terms involving only non-negative powers of z which will be of the form

$$e(M,\mathbf{x})\Sigma z^n - h(z),$$

the first sum being from 0 to $+\infty$ and the second term being a polynomial in z. Hence we obtain

$$P(h,M,z) = [(1-z^{w(1)})\dots(1-z^{w(d)})(1-z)]^{-1}[e(\mathbf{x},M) - (1-z)h(z)].$$

Now let z increase through real values to 1. Then the limit of $(1-z)^{d+1}P(h,M,z)$ is $e(\mathbf{x},M)/w(1)\dots w(d)$ which is a positive rational number. Further, this value depends only on M, the sequence of elements $x_1,\dots,x_d$ and the weights $w(1),\dots,w(d)$ and not on h. But $(1-z)^{d+1}P(h,M,z)$ does not depend on the choice of the basic reduction b(x) and hence not on either $x_1,\dots,x_d$ or on the weights $w(1),\dots,w(d)$, but only on the equivalence class of f.

We therefore denote $e(\mathbf{x},M)/w(1)\dots w(d)$ by e(f,M) and call this the *multiplicity function* of the filtration f. We now conclude this section by listing a number of properties of this function, this being little more than a rewriting of such properties of the ordinary multiplicity function e(x,M), for which we refer the reader to Northcott[LRMM].

THEOREM 8.2 i). *The function* e(f,M) *is an additive function on the category of finitely generated* Q*-modules and its values are non-negative rational numbers.*

ii). *If* f, g *are two noether filtrations on* Q *such that* $g(x) \geq f(x)$ *for all* x, *then* $e(g,M) \leq e(f,M)$ *for all* M.

iii). *If* rf *denotes the filtration obtained by multiplying the value of* f(x) *by the integer* r, *then*

$$e(rf,M) = r^{-d}e(f,M).$$

iv). e(f,M) *is determined by the values of* $e(f,Q/\mathbf{p})$, *where* $\mathbf{p}$ *ranges over those minimal prime ideals* $\mathbf{p}$ *of* Q *such that* $\dim Q/\mathbf{p} = d$.

v). *Let* f *be a basic filtration generated by* $x_1,\dots,x_d$ *with weights* $w(1),\dots,w(d)$. *Then if* f' *is the filtration* f/x_dQ *on* $Q' = Q/x_dQ$,

$$e(f,M) = (e(f',M/x_dM) - e(f',(0:x_d)_M))/w(d).$$

i) follows from the definition of e(f,M) as $e(\mathbf{x},M)/w(1)\dots w(d)$ and the corresponding properties of $e(\mathbf{x},M)$.

To prove ii) we note that, since $g(x) \geq f(x)$ for all x, it follows that

$$L(M/I_n(g)M) \leq L(M/I_n(f)M)$$

for all n. Hence the coefficients of $P(h,M,z) - P(h',M,Z)$ are all non-negative, where h is the good f-filtration on M defined by the sequence of sub-modules $I_n(f)M$ and h' the good g-filtration defined by the sequence $I_n(g)M$. It follows that the limit of $(1-z)^{d+1}(P(h,M,z) - P(h',M,z))$ as z increases to 1 is ≥ 0 and hence $e(f,M) \geq e(g,M)$.

iii) is immediate if we note that the result of multiplying f by r is to multiply each weight w(i) by r.

Since e(f,M) is an additive function on the category of finitely generated Q-modules and takes non-negative values, it follows that, if $\lambda(\mathbf{p},M)$ denotes the length of the $Q_{\mathbf{p}}$-module $M_{\mathbf{p}}$, where $\mathbf{p}$ is a minimal prime ideal of Q, then

$$e(f,M) = \Sigma\lambda(\mathbf{p},M)\, e(f,Q/\mathbf{p}).$$

Finally we note that $e(\mathbf{x},Q/\mathbf{p}) = 0$ if $\dim Q/\mathbf{p} < d$. Hence iv) is true.

For the proof of v), we note that it is a consequence of the corresponding result,

$$e(x_1,\ldots,x_d; M) = e(x'_1,\ldots,x'_{d-1}; M/x_dM) - e(x'_1,\ldots,x'_{d-1}; (0:x_d)_M).$$

This is the basis of the inductive definition given in Northcott[LRMM], on p.300 as (7.42) and its equivalence to the definition in terms of Koszul Complexes is given as theorem 5 on p.370 of the same reference.

9. THE DEGREE FUNCTION OF A NOETHER FILTRATION

1. Definition and elementary properties.

DEFINITION. *Let* $(Q,k,\mathfrak{m},d)$ *be a local ring,* x *be an element of* Q *satisfying the condition that* $\dim Q/xQ = d - 1$. *Let* f *be a noether filtration on* Q *with radical* $\mathfrak{m}$, *and let* M *be a finitely generated* Q*-module. Then, if* f_x *is the filtration* f/xQ *on* $Q_x = Q/xQ$, *we define* $d(f,M,x) = e(f_x,M/xM) - e(f_x,(0:x)_M)$.

We develop the elementary properties of $d(f,M,x)$, which we will refer to as the *degree function* on Q with respect to f, in a series of lemmas.

LEMMA 9.11. *If* f,x *are fixed,* $d(f,M,x)$ *is an additive function on the category of finitely generated* Q*-modules, taking non-negative rational values. Further, if* $\mathfrak{p}$ *is a prime ideal of* Q *such that* $\dim Q/\mathfrak{p} < d$, *then*

$$d(f,Q/\mathfrak{p},x) = 0.$$

Let

$$0 \longrightarrow L \longrightarrow M \longrightarrow N \longrightarrow 0$$

be an exact sequence of Q-modules and let $K_x(M)$ denote the Koszul Complex defined by the set consisting of the single element x. Then the homology modules of this complex are $H_0(M) = M/xM$, $H_1(M) = (0:x)_M$. Hence we have an exact sequence

$$0 \longrightarrow (0:x)_L \longrightarrow (0:x)_M \longrightarrow (0:x)_N \longrightarrow L/xL \longrightarrow M/xM \longrightarrow N/xN \longrightarrow 0$$

and therefore, if $a(M')$ is any additive function on the category of finitely generated Q_x-modules, it follows that $a(M,x) = a(M/xM) - a((0:x)_M)$ is an additive function on the category of Q-modules. We now prove that, if $a(M)$ is non-negative, then so is $a(M,x)$. It is sufficient to prove this if $M = Q/\mathfrak{p}$ where $\mathfrak{p}$ is a prime ideal of Q. Suppose x belongs to $\mathfrak{p}$. Then $M/xM = Q/\mathfrak{p} = (0:x)_M$. Hence $a(Q/\mathfrak{p},x) = 0$. If x does not belong to $\mathfrak{p}$, then $(0:x)_{A/\mathfrak{p}} = 0$, and hence $a(Q/\mathfrak{p},x) = a(Q/(\mathfrak{p} + xQ)) \geq 0$ since a is non-negative on the category of Q_x-modules.

Applying the above to $a(M) = e(f_x,M)$ on the category of finitely generated Q_x-modules, it follows that $d(f,M,x)$ is an additive, non-negative, rational-valued function on the category of finitely generated Q-modules, and hence by Lemma 1.33, that, if $\mathbf{p}$ is a non-minimal prime ideal of Q, then $d(f,Q/\mathbf{p},x) = 0$. If $\mathbf{p}$ is a minimal prime ideal of Q such that $\dim Q/\mathbf{p} < d$, and x belongs to $\mathbf{p}$, then we have seen above that $d(f,Q/\mathbf{p},x) = 0$. If, on the other hand, x does not belong to $\mathbf{p}$, then $d(f,Q/\mathbf{p},x) = e(f_x,Q/(\mathbf{p} + xQ))$ and $Q/(\mathbf{p} + xQ)$ has dimension $<d-1 = \dim Q_x$. Hence by the proof of 8.3(iii), $e(f_x,Q/(\mathbf{p} + xQ)) = 0$.

LEMMA 9.12. *If* f, M *are fixed, and* x_1,x_2 *are two elements of* Q *neither contained in any prime ideal* $\mathbf{p}$ *of* Q *such that* $\dim Q/\mathbf{p} = d$, *then*

$$d(f,M,x_1x_2) = d(f,M,x_1) + d(f,M,x_2).$$

It is sufficient to prove this in the case where $M = Q/\mathbf{p}$ where $\mathbf{p}$ is a prime ideal of Q such that $\dim Q/\mathbf{p} = d$. Then none of x_1, x_2, x_1x_2 belongs to $\mathbf{p}$, and hence we have $d(f,M,x) = e(f_x,M/xM)$ if $M = Q/\mathbf{p}$ and $x = x_1$, x_2 or x_1x_2. Now we have an exact sequence

$$0 \longrightarrow M/x_1M \xrightarrow{x_2} M/x_1x_2M \longrightarrow M/x_2M \longrightarrow 0.$$

and hence

$$e(f/x_1x_2Q, M/x_1x_2M) = e(f/x_1x_2Q, M/x_1M) + e(f/x_1x_2Q, M/x_2M).$$

But M/x_iM is annihilated by x_i, $i = 1,2$, and hence the filtration defined on M/x_iM by the sequence of sub-modules $I_n(f)(M/x_iM)$ is both an f/x_iQ-filtration and an f/x_1x_2Q-filtration. Hence $e(f/x_1x_2Q,M/x_iM) = e(f/x_iQ ,M/x_iM)$, $i = 1,2$, and the result follows.

LEMMA 9.13. *Let* f,g *be two noether filtrations on* Q *with radical* $\mathbf{m}$.

i). *If* f,g *are equivalent, then*

$$d(f,M,x) = d(g,M,x) \text{ for all } M,\ x.$$

ii). *If* f,g *satisfy* $f(y) \geq g(y)$ *for all* y *in* Q, *then*

$$d(f,M,x) \leq d(g,M,x) \text{ for all } M,\ x.$$

i) Follows as for the similar statement for $e(f,M)$, $e(g,M)$ since if f,g are equivalent so are f_x, g_x.

ii) Again we may assume that $M = Q/\mathbf{p}$ where $\dim Q/\mathbf{p} = d$. Then since, for all z in Q_x, $f_x(z) \geq g_x(z)$,

$$d(f,M,x) = e(f_x,M/xM) \leq e(g_x,M/xM) = d(g,M,x).$$

2. The degree formula: generalities.

Our purpose in the following sections is to prove a formula relating the additive function $d(f,M,x)$ to a sub-set of the set of valuations associated with f. This formula in its complete form is somewhat complicated, and we will commence by explaining the various terms which enter into it. First we will say that an ideal valuation v with centre $\mathfrak{m}$ on Q is a *good* valuation if the transcendence degree of its residue field K_v over k is d-1 (cf. Lemma 6.1). Note that we do not have to assume that v is an ideal valuation, providing it is integer-valued, since, by Theorem 6.24 ii), this follows from the statement about the transcendence degree. Further, the ideal $\mathbf{p}$ consisting of all elements x such that $v(x) = \infty$ is then a minimal prime ideal of Q such that $\dim Q/\mathbf{p} = d$. Now suppose M is any finitely generated Q-module and let

$$M = M_0 \supset M_1 \supset .... \supset M_N = (0)$$

be a filtration of M in which $M_{i-1}/M_i = Q/\mathbf{p}_i$ for some prime ideal $\mathbf{p}_i$ of Q. We recall that if $\mathbf{p}$ is a minimal prime ideal of Q, then the number of values of i for which $\mathbf{p}_i = \mathbf{p}$ is independent of the choice of the filtration. If $\mathbf{p}$ is the minimal prime ideal on which a given ideal valuation v takes the value ∞, then we will write this number as $L_v(M)$ or $L_{\mathbf{p}}(M)$, according as to whether we are primarily concerned with v or $\mathbf{p}$.

Next, consider the completion $Q^\wedge$ of Q and the completion $M^\wedge = M \otimes_Q Q^\wedge$ of M. First we recall that the filtration f has a natural extension $f^\wedge$ to $Q^\wedge$ which is also a noether filtration. Further, the valuation v considered above similarly has an extension $v^\wedge$ to $Q^\wedge$, and it takes the value ∞ on a prime ideal $\mathbf{p}^\wedge$ which is a minimal prime ideal of $\mathbf{p}Q^\wedge$. We denote by $\delta(v)$ or $\delta(\mathbf{p}^\wedge)$ the length of the isolated primary component of $\mathbf{p}Q^\wedge$ with radical $\mathbf{p}^\wedge$. Next consider $M^\wedge$. The filtration of M given above yields a filtration

$$M^\wedge = M^\wedge_0 \supset M^\wedge_1 \supset ... \supset M^\wedge_N = (0)$$

where $M^\wedge_{i-1}/M^\wedge_i = Q^\wedge/\mathbf{p}_i Q^\wedge$. Then $Q^\wedge/\mathbf{p}_i Q^\wedge$ will have a filtration with $\delta(\mathbf{p}^\wedge)$ quotients isomorphic to $Q^\wedge/\mathbf{p}^\wedge$, for each minimal prime ideal $\mathbf{p}^\wedge$ of $\mathbf{p}Q^\wedge$, and, for each such prime ideal, $L_{\mathbf{p}^\wedge}(M^\wedge) = \delta(\mathbf{p}^\wedge) L_{\mathbf{p}}(M)$.

Now consider the function $d(f,M,x)$ as a non-negative additive function on the category of finitely generated Q-modules. As such, it is determined by the values of $d(f,Q/\mathbf{p},x)$, where $\mathbf{p}$ ranges over the minimal prime ideals of Q, and is given by

$$d(f,M,x) = \Sigma L_{\mathbf{p}}(M)d(f,Q/\mathbf{p},x).$$

Next we turn to the function $d(f\hat{},M,x)$, where $f\hat{}$ is the canonical extension of f to $Q\hat{}$ and the function is considered as a non-negative additive function on the category of finitely generated $Q\hat{}$-modules. From the definition of $d(f,M,x)$, it is clear that, if x belongs to Q and M is a finitely generated Q-module, then,

$$d(f,M,x) = d(f\hat{},M\hat{},x) = \Sigma L_{\mathbf{p}\hat{}}(M)d(f\hat{},Q\hat{}/\mathbf{p}\hat{},x) = \Sigma\delta(\mathbf{p}\hat{})L_{\mathbf{p}}(M)d(f\hat{},Q\hat{}/\mathbf{p}\hat{},x)$$

$$= \Sigma\delta(\mathbf{p}\hat{})L_{\mathbf{p}}(M)d(f\hat{}/\mathbf{p}\hat{},Q\hat{}/\mathbf{p}\hat{},x) \qquad (9.21)$$

where $\mathbf{p}\hat{}$ ranges over the minimal prime ideals of $Q\hat{}$, and, as above $\mathbf{p} = Q\cap\mathbf{p}\hat{}$.

We can now state a preliminary form of the degree formula.

If $(Q,\mathfrak{m},k,d)$ *is a local ring and* f *is a noether filtration on* Q *with radical* $\mathfrak{m}$, *then*

$$d(f,M,x) = \Sigma d(f,M,v)v(x)$$

where $d(f,M,v)$ *is defined for all good ideal valuations on* Q *centre* $\mathfrak{m}$, *and is a non-negative, rational-valued additive function on the category of finitely generated* Q*-modules. Further,* $d(f,M,v)$ *is zero for all* M *if* v *is not associated with* f.

It is clear from the above that, in order to prove this, it will be sufficient to prove that $d(f\hat{}/\mathbf{p}\hat{},Q\hat{}/\mathbf{p}\hat{},x)$ can be written in the form $\Sigma d(f\hat{}/\mathbf{p}\hat{},v\hat{})v\hat{}(x)$, the sum being over all good ideal valuations on $Q\hat{}/\mathbf{p}\hat{}$, with centre the maximal ideal $\mathfrak{m}\hat{}/\mathbf{p}\hat{}$ of $Q\hat{}/\mathbf{p}\hat{}$. This essentially means that we can restrict attention to the case where Q is a complete local domain. In fact, we do not need to make such a strong restriction. We will assume therefore only that Q is i) a domain, ii) analytically unramified and iii) quasi-unmixed. This we do in the next section.

3. The degree formula: preliminary form.

Our object in this section is the following theorem.

THEOREM 9.31. *Let* $(Q,\mathfrak{m},k,d)$ *be a local domain which is i) analytically unramified, ii) quasi-unmixed. Let* f *be a noether filtration on* Q *with centre* $\mathfrak{m}$, *and let* $d(f,x)$ *denote* $d(f,Q,x)$. *Then*

$$d(f,x) = \Sigma d'(f,v)v(x)$$

where v *ranges over the finite set of valuations associated with* f *and* $d'(f,v)$ *is a positive rational number.*

We will have to prove a number of preliminary results before we can prove this theorem. These we will state as lemmas. We recall that Q_n denotes the localisation of $Q[X]$ at $\mathfrak{m}[X]$, where X denotes a set of n independent indeterminates over Q.

LEMMA 9.32. *Let* $(Q,\mathfrak{m},k,d)$ *be a local domain, and let* $J = (x_1,\ldots,x_d)$ *be an* $\mathfrak{m}$*-primary ideal of* Q. *Let* $s \leq d-1$, $B(s)$ *be the ring* $Q[x_1/x_d,\ldots,x_s/x_d]$, $X_1,\ldots,X_{d-1}$ *be indeterminates over* Q, *and let* $\mathfrak{P}_s$ *be the kernel of the homomorphism of* $Q[X_1,\ldots,X_s]$ *onto* $B(s)$ *in which* X_i *is mapped onto* x_i/x_d. *Then, for all sufficiently large* N,

$$\mathfrak{P}_s = \mathfrak{a}_s : J_s^N$$

where $\mathfrak{a}_s = (x_dX_1-x_1,\ldots,x_dX_s-x_s)$ *and* J_s *is the ideal generated by* $(x_1,\ldots,x_s,x_d)$.

$\mathfrak{P}_s$ consists of all polynomials $f(X_1,\ldots,X_s)$ over Q such that

$$f(x_1/x_d,\ldots,x_s/x_d) = 0.$$

Such a polynomial can be written in the form $F(X_1,\ldots,X_s,1)$, where F is a homogeneous polynomial over Q in the indeterminates $X_1,\ldots,X_s,X_d$ such that $F(x_1,\ldots,x_s,x_d) = 0$. Suppose that F is such a polynomial and that its degree is n. Then

$$x_d^nF(X_1,\ldots,X_s,X_d) = F(x_dX_1,\ldots,x_dX_s,x_dX_d) =$$

$$F(x_1X_d + (x_dX_1 - x_1X_d),\ldots,x_sX_d + (x_dX_s - x_sX_d),x_dX_d) =$$

$$X_d^nF(x_1,\ldots,x_s,x_d) + \Sigma(x_dX_i - x_iX_d)G_i(X_1,\ldots,X_s,X_d),$$

the latter sum being from $i = 1$ to s, and the polynomials $G_i(X_1,\ldots,X_s,X_d)$ being homogeneous of degree n-1. Since the first term is zero it follows that x_d^nF belongs to the ideal $(x_dX_1 - x_1X_d,\ldots,x_dX_s - x_sX_d)$ and hence, putting $X_d = 1$, x_d^nf belongs to the ideal $\mathfrak{a}_s = (x_dX_1 - x_1,\ldots,x_dX_s - x_s)$. Then

$$x_i^nf = x_d^nX_i^nf - (x_d^nX_i^n - x_i^n)f$$

also belongs to $\mathfrak{a}_s$. Hence, $(x_1^n,\ldots,x_s^n,x_d^n)f$ is contained in $\mathfrak{a}_s$. This implies that, for each f, and for N sufficiently large (how large depending on f), J_s^Nf is contained in $\mathfrak{a}_s$. But as N increases, the ideals $\mathfrak{a}_s : J_s^N$ form an ascending sequence of ideals, and so $\mathfrak{a}_s : J_s^N$ is constant for large N. It follows that $\mathfrak{P}_s$ is equal to this constant value since, as Q is a domain, it is clear that $\mathfrak{P}_s$ contains $\mathfrak{a}_s : J_s^N$ for all N.

LEMMA 9.33. *Let* $(Q,\mathfrak{m},k,d)$, J_s, $B(s)$, $\mathfrak{a}_s$ *be as above. Then the ideal* $\mathfrak{m}B(s)$ *is prime, and the localisation* $L(s)$ *of* $B(s)$ *at* $\mathfrak{m}B(s)$ *is isomorphic to* $Q_s/\mathfrak{P}_sQ_s$, *where* Q_s *is the localisation of* $Q[X_1,\ldots,X_s]$ *at* $\mathfrak{m}[X_1,\ldots,X_s]$. *Further,* $L(s)$ *is a* $(d-s)$*-dimensional local domain, and, if* Q *is analytically unramified, so is* $L(s)$.

If $J = J_{d-1}$, J is $\mathfrak{m}$-primary and f_J (or J) has analytic spread d by Lemma 6.12. Hence the homomorphism of $k[X_1,\ldots,X_d]$ onto $F(f_J,\mathfrak{m})$, in which X_i maps to the image of x_it, is an isomorphism. It follows that any homogeneous polynomial $F(X_1,\ldots,X_d)$ over Q such that $F(x_1,\ldots,x_d) = 0$ has all its coefficients in $\mathfrak{m}$. Hence, for all s, the prime ideal $\mathfrak{P}_s$ of $Q[X_1,\ldots,X_s]$ is contained in $\mathfrak{m}Q[X_1,\ldots,X_s]$, implying that $B(s)/\mathfrak{m}B(s) = k[X_1,\ldots,X_s]$, and therefore that $\mathfrak{m}B(s)$ is prime. Further, the localisation $L(s)$ of $B(s)$ at $\mathfrak{m}B(s)$ is isomorphic to $Q_s/\mathfrak{P}_sQ_s$. Now, writing z_i for $x_dX_i - x_i$,

$$JQ_s = (z_1,\ldots,z_s,x_{s+1},\ldots,x_d).$$

It follows that $\mathfrak{a}_sQ_s = z_1Q_s+\ldots+z_sQ_s$ is generated by a sub-set of a set of parameters. Hence, by [Z-S] vol. 2, p.292, $Q_s/\mathfrak{a}_sQ_s$ has dimension d-s. Clearly $\mathfrak{a}_sQ_s$ has height s. Hence $L(s)$ also has dimension d-s, since $\mathfrak{a}_s$ has height s while J_s has height s+1, so that $\mathfrak{P}_s = \mathfrak{a}_s : J_s^N$ must be the sole prime ideal containing $\mathfrak{a}_s$ of height s and $Q_s/\mathfrak{P}_sQ_s$ dimension d-s.

If Q is analytically unramified, so is $L(s)$ as a localisation of a finitely generated extension of Q contained in the field of fractions of Q (see the Corollary to Theorem 5.41).

LEMMA 9.34. *With the notation of the last two lemmas, if* f *is the noether filtration* f_J, *then, for any non-zero* x *in* Q,

$$d(f,x) = e(\mathfrak{a}_{d-1}Q_{d-1} + xQ_{d-1}).$$

By definition, $d(f,x) = e((J + xQ)/xQ) = e((JQ_{d-1} + xQ_{d-1})/xQ_{d-1})$. Let Q' denote the local ring Q/xQ and let z' denote the image of an element z of Q in Q', with a similar notation for ideals. Then J' is generated by the elements $x'_1,\ldots,x'_d$. Now Q' has dimension d-1, implying that J' has analytic spread $\leq d-1$. Hence the graded map

$K[X_1,\dots,X_d] \to F(f_{J'},\boldsymbol{m})$ is not an isomorphism i.e., there exists a homogeneous polynomial $g(Y_1,\dots,Y_d)$ of degree N over Q' in indeterminates $Y_1,\dots,Y_d$ such that not all coefficients of g are in $\boldsymbol{m}Q'$ and $g(x'_1,\dots,x'_d) = 0$. Now, let $z_i = x_dX_i - x_i$, as in the last lemma. Then we can write the above as

$$g(x'_dX_1 - z'_1,\dots,x'_dX_{d-1} - z'_{d-1},x'_d) = 0$$

in which, when we expand the left-hand side as a homogeneous polynomial over Q'_{d-1} in $z'_1,\dots,z'_{d-1},x'_d$, the coefficient of x'^N_d is $g(X_1,\dots,X_{d-1},1)$ and so is a unit of Q'_{d-1}. Hence we have proved that $(JQ_{d-1}+xQ_{d-1})/xQ_{d-1}$ is contained in the integral closure of $(\boldsymbol{a}_{d-1} + xQ_{d-1})/xQ_{d-1}$, and so these two ideals have the same integral closure and therefore the same multiplicity. Hence

$$d(f,x) = e(\,(xQ_{d-1}+\boldsymbol{a}_{d-1})/xQ_{d-1}).$$

Now consider the ideal $xQ_{d-1} + \boldsymbol{a}_{d-1}$ of Q_{d-1}. This is generated by the d elements $z_1,\dots,z_{d-1},x$, and x is not a zero divisor. Hence, by the formula of Northcott and Wright used in the proof of 8.2(v) and quoted on p.129, it follows that

$$e(xQ_{d-1}+\boldsymbol{a}_{d-1}) = e((xQ_{d-1}+\boldsymbol{a}_{d-1})/xQ_{d-1}) = d(f,x)$$

as required.

LEMMA 9.35. *With the notation of the previous lemmas, if* f *is the noether filtration* f_J *on* Q, *and* x *is a non-zero element of* Q,

$$d(f,x) = l(L(d-1)/xL(d-1)).$$

We prove this by induction on d. If d = 1, then L(0) = Q and the statement simply reduces to the statement that the multiplicity of an ideal in a local ring of dimension 0 is the length of the ring. Now suppose that Q has dimension d > 1. Then L(1) is a local ring of dimension d-1. Further, JL(1) is generated by $x_2,\dots,x_d$, and finally, if we write L(s) as L(s,Q), to indicate its dependence on Q, then L(d-1,Q) = L(d-2,L(1)). Now let fL(1) denote the filtration on L(1) determined by the sequence of ideals $I_n(f)L(1)$. Then applying our inductive hypothesis, we have $d(fL(1),x) = l(L(d-1)/xL(d-1))$. Now we recall that d(f,x) is equal to the multiplicity of the ideal $(z_1,\dots,z_{d-1},x)$ of Q_{d-1}. Since the latter is a domain, this is equal to the multiplicity of the ideal $(z'_2,\ldots,z'_{d-1},x')$ of $Q' = Q_{d-1}/z_1Q_{d-1}$. But we can identify $L(1)(d-2)$ with the localisation of $L(1)[X_2,\ldots,X_{d-1}]$ at $\mathfrak{m}L(1)[X_2,\ldots,X_{d-1}]$, and so we have a homomorphism of Q' onto L(1)(d-2) whose kernel is annihilated by (x_1,x_2) by Lemma 9.32. Hence d(f,x) is equal to the multiplicity of the ideal $(z_2,\ldots,z_{d-1},x)$ of L(1),and by the last lemma this is equal to d(fL(1),x), so finally we have

$$d(f,x) = l(L(d-1)/xL(d-1)).$$

LEMMA 9.36. *Let* $(Q,\mathfrak{m},k,d)$ *be a local domain which is both analytically unramified and quasi-unmixed, let* $J = (x_1,\ldots,x_d)$ *be an* $\mathfrak{m}$*-primary ideal of* Q, *and let* $f = f_J$. *Then there exist integer-valued valuations* $v_1,\ldots,v_s$ *on the field of fractions* F *of* Q *such that*

i) $v_i(x) \geq 0$ *on* Q *and* >0 *on* $\mathfrak{m}$, $i = 1,\ldots,s$,

ii) *for each* i, *the elements* x_j/x_d, $j = 1,\ldots,d-1$, *belong to the valuation ring* O_i *of* v_i, *and, if* $y_{i,j}$ *denotes the image of* x_j/x_d *in the residue field* K_i *of* v_i, *then* $y_{i,1},\ldots,y_{i,d-1}$ *form a transcendence basis of* K_i *over* k,

iii) $d(f,x) = \Sigma[K_i : k(y_{i,1},\ldots,y_{i,d-1})]v_i(x)$, *the sum being for* $i = 1$ *to* s.

Finally, $v_1,\ldots,v_s$ *are the valuations associated with* f *(or with the ideal* J).

We commence by observing that, as Q is analytically unramified, so is L = L(d-1) by the Corollary to Theorem 5.41. This implies that the integral closure L* of L is a finite L-module and hence that L*/L is an L-module of finite length. Now, by the Krull-Akizuki Theorem, L* is a semi-local ring of dimension 1, with maximal ideals $\mathfrak{n}_1,\ldots,\mathfrak{n}_s$ such that the localisation of L* at $\mathfrak{n}_i$ is a discrete valuation ring O_i. We take $v_1,\ldots,v_s$ to be the normalised valuations associated with the rings $O_1,\ldots,O_s$, so that i) is satisfied. Now the residue field K of L is isomorphic to a pure transcendental extension of k of trans.deg. d-1, the images of x_j/x_d, $j = 1,\ldots,d-1$, forming a transcendence basis of K over k. Further, K_i is a finite algebraic extension of K for each i. Hence ii) is satisfied. To prove iii), we note that, if l(.) denotes lengths of L-modules,

$$l(L/xL) = e(x,L) = e(x,L^*) = l(L^*/xL^*) = \Sigma[K_i : K]v_i(x)$$

the sum being from i = 1 to s, which is the statement of iii).

To finish the proof, we have to identify the valuations $v_1,\dots,v_s$ with the associated valuations of $f = f_J$.

Form the ring $G(f) = Q[tx_1,\dots,tx_d,u]$. Since $J = (x_1,\dots,x_d)$ is $\mathfrak{m}$-primary, $s(f) = d$ and it follows that $G(f)/(uG(f) + \mathfrak{m}G(f))$ is isomorphic to $k[X_1,\dots,X_d]$. Hence $uG(f)$ has radical the prime ideal $\mathfrak{P} = uG(f) + \mathfrak{m}G(f)$. Consider a valuation v on the field of fractions F of Q which is associated with f. The valuation v is the restriction to F of a valuation V associated with the filtration f_u on G(f). Now V is a Krull valuation of G(f). Let its centre on G(f) be $\mathfrak{P}'$. Then $\mathfrak{P}' \supseteq \mathfrak{P}$. Let $\mathfrak{L}'$ be the localisation of G(f) at $\mathfrak{P}'$. Since Q is quasi-unmixed, so is $\mathfrak{L}'$ by Theorem 7.32 (which implies that $\mathfrak{L}'$ has only one quasi-height). Now V is an ideal valuation of $\mathfrak{L}'$ by Theorem 7.24 and has as its centre on $\mathfrak{L}'$ the maximal ideal of $\mathfrak{L}'$. Hence its residue field must have transcendence degree equal to $\dim \mathfrak{L}' - 1$ over the residue field of $\mathfrak{L}'$ But V being a Krull Valuation of $\mathfrak{L}'$, its residue field is a finite algebraic extension of the residue field of $\mathfrak{L}'$ by Theorem 3.21. Hence $\mathfrak{P}'$ has height 1, and so equals $\mathfrak{P}$ and hence we can write $\mathfrak{L}$ in place of $\mathfrak{L}'$. Then $\mathfrak{L}$ contains the elements x_i/x_d and hence the ring B(d-1). The centre of V on B(d-1) will be $\mathfrak{m}B(d-1)$ and hence $V(x) \geq 0$ on $L = L(d-1)$, and $v(x) \geq 0$ on L(d-1). Hence $v(x) \geq 0$ on L and therefore on L^*, implying that v is one of the valuations $v_1,\dots,v_s$.

Now let w be one of the valuations v_i. Then w has a graded extension W to $L[u,x_dt]$ obtained by defining W(u) to be $w(x_d)$, so that $W(tx_d) = 0$. Now $L[u,xt_d]$ contains $tx_j = (x_j/x_d).tx_d$ for each j and so contains G(f). Consider the centre of W on G(f). It contains $\mathfrak{P}$ since it contains u and $\mathfrak{P}$ contains $\mathfrak{m}G(f)$. If it is not $\mathfrak{P}$, then it would contain an element of positive degree not in $\mathfrak{m}G(f)$, i.e. an element of the form $z = g(tx_1,\dots,tx_d)$ where $g(Y_1,\dots,Y_d)$ is a homogeneous polynomial, of degree N say, whose coefficients do not all belong to $\mathfrak{m}$. But then

$$W(g(x_1/x_d,\dots,x_{d-1}/x_d,1)) = W((tx_d)^{-N}.z) = W(z) > 0.$$

But $g(x_1/x_d,\dots,x_{d-1}/x_d,1)$ is an element of B(d-1) and since

$$B(d-1)/\mathfrak{m}\, B(d-1) = k[X_1,\dots,X_{d-1}]$$

the condition on g implies that g does not belong to $\mathfrak{m}$ B(d-1), which is the centre of w on B(d-1). Hence we have a contradiction, and W must have centre $\mathfrak{P}$ of height 1 on

$G(f)$ and hence is a Krull valuation of $G(f)$ such that $W(u) > 0$. Hence it is associated with the filtration $f_{uG(f)}$ and therefore w is associated with f.

LEMMA 9.37 i). *Let* $(Q,\mathfrak{m},k,d)$ *be a local ring and let* f *be a noether filtration on* Q *of analytic spread* s. *Then there is an integer* w *such that, for any integer* k, *there is a basic filtration* b_k *equivalent to* f *and* b_k *is generated by elements* $x_1,\ldots,x_s$ *all with weight* kw.

ii). *Let* J *be an ideal of* Q *of analytic spread* s. *Then there exists an integer* w *such that* J^{kw} *has a reduction* B *generated by* s *elements for all integers* k.

i) Let f be as given. Then we can replace f by an equivalent filtration g generated by s elements $y_1,\ldots,y_s$ with weights $w(1),\ldots,w(s)$ by Theorem 6.12. Let w be the l.c.m. of $w(1),\ldots,w(s)$. Then the filtration b_k generated by the elements $x_i = y_i^{kw/w(i)}$, each taken with weight kw, is also equivalent to f.

ii) Take f to be f_J. Then, if b_k is as defined above, and $n = qkw + r$ with $0 \le r < kw$, and B is the ideal generated by $x_1,\ldots,x_s$,

$$I_n(b_k) = B^q.$$

But, as b_k is a reduction of f_J, then, by the remark on p.87,

$$J^{n+kw} = J^n B$$

for all large n. Choose n of the form kmw for m sufficiently large. Then

$$(J^{kw})^{m+1} = (J^{kw})^m.B$$

and B is a reduction of J^{kw}.

We are now in a position to prove Theorem 9.31.

Proof of Theorem 9.31. We first note that, as f has radical $\mathfrak{m}$, $s(f) = d$. Then we can replace f by the equivalent filtration b_1 obtained by putting $k = 1$ in the last lemma. Then b_1 takes values multiples of w. Replace it by the filtration f_B, where B is the ideal generated by $x_1,\ldots,x_d$. Then b_1 and hence f are equivalent to $w.f_B$. Hence

$$d(f,x) = d(w.f_B, x) = w^{-(d-1)}.d(f_B,x)$$

by Theorem 8.2 iii) and the definition of degree function. Hence, by Lemma 9.36,

$$d(f,x) = \Sigma\,[K_i : k(y_{i,1},\ldots,y_{i,d-1})]v_i(x)/w^{d-1}$$

which is an expression of the required form.

4. The degree formula: final version.

We are now in a position to state the final version of the degree formula.

THEOREM 9.41. *Let* $(Q,\mathfrak{m},k,d)$ *be a local ring. Let* v *range over the set* Φ *of good ideal valuations on* Q *with centre* $\mathfrak{m}$. *Let* f *be a noether filtration on* Q *centre* $\mathfrak{m}$. *Then there exists a function* $d(f,v)$ *on the set* Φ *which takes non-negative rational values, and which is equal to zero if* v *is not one of the valuations associated with* f, *such that*

$$d(f,M,x) = \Sigma_v\, \delta(v)L_v(M)d(f,v)v(x)$$

where $\delta(v)$, $L_v(M)$ *have the meaning defined at the beginning of the last section.*

We commence by referring back to the formula (9.21)

$$d(f,M,x) = \Sigma\delta(\mathbf{p}^{\wedge})L_{\mathbf{p}}(M)d(f^{\wedge}/\mathbf{p}^{\wedge},Q^{\wedge}/\mathbf{p}^{\wedge},x)$$

where the sum is over the minimal prime ideals $\mathbf{p}^{\wedge}$ of $Q^{\wedge}$ and $\mathbf{p} = \mathbf{p}^{\wedge}\cap Q$. We now apply Theorem 9.31 to write

$$d(f^{\wedge}/\mathbf{p}^{\wedge},Q^{\wedge}/\mathbf{p}^{\wedge},x) = \Sigma d(f^{\wedge}/\mathbf{p}^{\wedge},v)v(x)$$

where v ranges over all good ideal valuations v whose extension to $Q^{\wedge}$ takes the value ∞ on $\mathbf{p}^{\wedge}$. This is possible, since the ring $Q^{\wedge}/\mathbf{p}^{\wedge}$ is a complete local domain, and hence is both analytically unramified and quasi-unmixed. It follows that $\mathbf{p}^{\wedge}$ is determined by v, and hence $f^{\wedge}/\mathbf{p}^{\wedge}$ is determined by f and v. Hence we can write $d(f,v)$ in place of $d(f^{\wedge}/\mathbf{p}^{\wedge},v)$. Further, we note that $d(f^{\wedge}/\mathbf{p}^{\wedge},v)$ is zero save when $v^{\wedge}$ is associated with $f^{\wedge}/\mathbf{p}^{\wedge}$, i.e., except when v is associated with f. Finally, we can write $\delta(v)$ in place of $\delta(\mathbf{p}^{\wedge})$ and $L_v(M)$ in place of $L_{\mathbf{p}}(M)$, and so obtain the final formula

$$d(f,M,x) = \Sigma\delta(v)L_v(M)d(f,v)v(x).$$

It should be emphasised that the degree formula is essentially an existence formula, in that it implies the existence of non-negative rational numbers $d(f,v)$ such that the above formula is true. It is therefore natural to ask if these numbers not only exist, but are unique. This is the point of the next theorem.

THEOREM 9.42. *Let* A *be a noetherian ring. Let* $\mathbf{p}_1,\dots,\mathbf{p}_s$ *be the minimal prime ideals of* A *and let* F_i *denote the field of fractions of* $A/\mathbf{p}_i$. *Let* S *be a finite set of functions* $\{w_{ij}\}$, $i = 1,\dots,s$; $j = 1,\dots,r_i$, *on* A *taking integral values together with* ∞, *each* w_i *being of the form* $v_{ij}(f_i(x))$ *where* f_i *is the homomorphism of* A *into* F_i, *and* v_{ij} *is an integer valued valuation on* F_i. *Assume further that the valuations* v_{ij} *have distinct valuation rings, and each takes a value distinct from* $0,\infty$. *Then, if real numbers* a_{ij} *satisfy*

$$\Sigma a_{ij} w_{ij}(x) = 0$$

for all x *for which* $w_{ij}(x)$ *is finite for all* i,j, $a_{ij} = 0$ *for all* i,j.

We first introduce some simplifications. Let $\mathfrak{r}$ be the radical of A. Then we can replace A by $A/\mathfrak{r}$, i.e., we can assume that A has no nilpotent elements. The restriction on x is then that it be a non-zero divisor of A. We can now replace A by its complete ring of fractions which is the direct sum of fields $F_1 \oplus \ldots \oplus F_s$ and replace the valuations v_{ij} by their unique extensions to this ring. The equation

$$\Sigma a_{ij} w_{ij}(x) = 0$$

remains valid for all non-zero divisors of this extension.

We are thus reduced to the case where A is a direct sum of fields. Fix a valuation v_{kl}. Then we can choose an element x_k of F_k such that $v_{kl}(x_k) = 1$ and $v_{kj}(x_k) = 0$ if $j \neq l$. Further, if $i \neq k$ we can choose x_i in F_i such that $v_{ij}(x_i) = 0$ for all j. Let x be the element Σx_i. Then it is clear that $w_{kl}(x) = 1$ and $w_{ij}(x) = 0$ if either $i \neq k$ or $j \neq l$. Hence $a_{kl} = 0$.

COROLLARY. *Let A be a local ring* $(Q,\mathfrak{m},k,d)$, *let* f *be a noether filtration on* Q *with radical* $\mathfrak{m}$ *and let the functions* $w_{ij}(x)$ *be the valuations associated with* f. *Then if* b_{ij} *are rational numbers such that*

$$d(f,x) = \Sigma b_{ij} w_{ij}(x)$$

for all x *such that* $w_{ij}(x)$ *is finite for all* i,j, *then*

$$b_{ij} = \delta(w_{ij}) d(f,w_{ij}).$$

This is an immediate consequence of the formula

$$d(f,x) = \Sigma\, \delta(w_{ij}) d(f,w_{ij}) w_{ij}(x)$$

and the theorem above.

We conclude this chapter with an application of the degree formula. First we require a lemma.

LEMMA 9.43. *Let* $(Q,\mathfrak{m},k,d)$ *be a local ring, and let* J *be an* $\mathfrak{m}$*-primary ideal of* Q. *For any valuation* $v \geq 0$ *on* Q, *and* >0 *on* $\mathfrak{m}$, *let* v(J) *be the minimum value of* v(x) *on* J. *Finally, write* d(J,x) *for* $d(f_J,x)$ *and* d(J,v) *for* $d(f_J,v)$. *Then*

$$e(J) = \Sigma\delta(v)L_v(Q)d(J,v)v(J)$$

If, in Theorem 9.41, we take $f = f_J$ and M = Q, we obtain

$$d(J,x) = \Sigma\delta(v)L_v(Q)d(J,v)v(x)$$

with the notation introduced above. Next, replacing J by J^r, we have

$$d(J^r,x) = r^{d-1}d(J,x),$$

and the Corollary to Theorem 9.42 now implies that $d(J^r,v) = r^{d-1}d(J,v)$ for all r and v. Now we turn to the result to be proved. Since $e(J^r) = r^d e(J)$, and $v(J^r) = rv(J)$, it is clear that, if we can prove the result for some power of J, it will follow for J. We use this fact to require that J has a basic reduction B, since a sufficiently high power of any ideal has a basic reduction by Northcott and Rees[1954b]. Finally, we note that, if B is a reduction of J, then d(B,x) = d(J,x), which implies, by the Corollary to Theorem 9.42, that d(J,v) = d(B,v) for all v. Since e(J) = e(B) and v(J) = v(B) for all v, we may finally impose the condition that J is generated by d elements $x_1,\ldots,x_d$.

In this case, since the valuations v for which d(f,v) are non-zero are ≥ 0 on L(d-1), with the notation of Lemmas 9.33-9.36, and $JL(d-1) = x_dL(d-1)$, it follows that, for such a valuation, $v(J) = v(x_d)$. Then, by Lemma 9.34,

$$d(J,x_d) = e((x_dX_1 - x_1 ,\ldots, x_dX_{d-1} - x_{d-1} , x_d)Q_{d-1}) = e(JQ_{d-1}) = e(J).$$

But

$$d(J,x_d) = \Sigma\delta(v)L_v(Q)d(J,v)v(x_d) = \Sigma\, \delta(v)L_v(Q)d(J,v)v(J)$$

proving the result.

THEOREM 9.44. *Let* $(Q,\mathfrak{m},k,d)$ *be a quasi-unmixed local ring and let* f,g *be two noether filtrations on* Q *with radical* $\mathfrak{m}$, *such that* $g(x) \geq f(x)$ *for all* x. *Then a necessary and sufficient condition for* f,g *to be equivalent is that*

$$e(f,Q) = e(g,Q).$$

The necessity is immediate.

To prove sufficiency, we first introduce the restriction on Q that the field k is infinite.

The proof will be by a series of stages in which, at each stage, we modify the result to be proved.

First we note that we can choose an integer w such that f is equivalent to $w.f_J$ and g is equivalent to $w.f_K$ where $J = I_w(f)$, $K = I_w(g)$, are $\mathfrak{m}$-primary ideals of Q such that $K \supseteq J$. It is now sufficient to prove that f_J, f_K are equivalent under the assumptions that $K \supseteq J$ and $e(J) = e(K)$. Note that this follows since

$$e(J) = w^d e(f,Q) = w^d e(g,Q) = e(K).$$

Our result will follow if we can prove that K is in the integral closure of J.

Now suppose that we join K, J by a sequence of ideals

$$K = K_0 \supset K_1 \supset \dots \supset K_n = J,$$

with $l(K_i/K_{i+1}) = 1$. It then follows that all the ideals K_i have the same multiplicity, and it will be sufficient to prove that K_i is in the integral closure of K_{i+1} for each i. Hence we may further assume that $K = J + xQ$ for some x.

Now as k is infinite, J has a reduction B generated by d elements. Then $J^{r+1} = J^r B$ for all sufficiently large r and

$$K^{r+1} \supseteq (B + xQ)K^r \supseteq J^r B + xK^r = J^{r+1} + x(xQ + J)^r = (J + xQ)^{r+1} = K^{r+1}$$

and hence $B + xQ$ is a reduction of K. Hence replacing J, K, by B, $B + xQ$, we may further assume that J is generated by d elements.

Now let $v_1,\dots,v_k$ be the valuations associated with J, i.e., with f_J. Then, as Q is quasi-unmixed, we have $v_i(y) = v_i(J)$ for each i and each y which is in J but not in $\mathfrak{m}J$, and, further, $d(J,v) \neq 0$ if and only if v is one of $v_1,\dots,v_k$ by Lemma 9.36.

Further, to prove that K is in the integral closure of J, it will be sufficient to find an element x such that $K = J + xQ$ and $v_i(x) \geq v_i(J)$ for $i = 1,\dots,k$. We commence by proving that we can choose x to satisfy the opposite, i.e., that $v_i(x) \leq v_i(J)$ for $i = 1,\dots,k$. We can, without altering the fact that $K = J + xQ$, replace the given x by $x + y$, where y is any element of J. We now choose such a y, which we suppose is not in $J\mathfrak{m}$. First, if $v_i(x) < v_i(J)$, then, however y is chosen, $v_i(x+y) < v_i(J)$. If $v_i(x) > v_i(J)$, then, however y is chosen, $v_i(x+y) = v_i(J)$. Finally, if $v_i(x) = v_i(J)$, then we must choose y such that $v_i(x+y) = v_i(J)$. Since k is assumed infinite, we can choose such a y satisfying this condition for the finitely many i for which $v_i(x) = v_i(J)$.

Now consider $d(J,x)$, $d(K,x)$. Since $K = J + xQ$, these are equal. But as x belongs to

K, $d(K,x) \geq e(K,Q)$. Since $v_i(x) \leq v_i(J)$, $v(x) \geq v(J)$ if $d(J,v) \neq 0$. Hence

$$e(K,Q) \leq d(J,x) = \Sigma\delta(v)L_v(Q)d(J,v)v(x) \leq \Sigma\delta(v)L_v(Q)d(J,v)v(J) = e(J,Q)$$

and the equality $e(K,Q) = e(J,Q)$ implies that we must have $v_i(x) = v_i(J)$ for each i, and so the choice of x must satisfy $v_i(x) \geq v_i(J)$ for all i, that is, x is in the integral closure of J. It therefore follows that f_J, f_K are equivalent, and this in turn implies that f, g are equivalent.

We now remove the restriction that k is infinite. We replace Q by Q_1, and f,g by their natural extensions f_1,g_1 to Q_1. Then the residue field of Q_1 is infinite, and f_1,g_1 satisfy the conditions that $g_1(x) \geq f_1(x)$ for all x and $e(f_1,Q_1) = e(g_1,Q_1)$ by 7.4 and 8.1. Hence we can prove that f_1, g_1 are equivalent, and this clearly implies that their restrictions f,g to Q are equivalent.

10. THE GENERAL EXTENSION OF A LOCAL RING

1. Introduction.

Let $(Q,\mathfrak{m},k,d)$ be a local ring, M be a finitely generated Q-module, and let $X_1,X_2,\ldots$ be a countable set of indeterminates over Q. Then we define the general extensions Q_g, M_g of Q, M as follows.

DEFINITION. *The general extension* Q_g *of* Q *is the localisation of* $Q[X_1,X_2,\ldots]$ *at the prime ideal* $\mathfrak{m}[X_1,X_2,\ldots]$. *The general extension* M_g *of* M *is* $M\otimes_Q Q_g$.

The kernel of the map $Q[X_1,X_2,\ldots] \to Q_g$ is zero, since it consists of all polynomials in $X_1,X_2,\ldots$ annihilated by some polynomial g which has at least one coefficient a unit of Q. But then g is not a zero divisor of $Q[X_1,X_2,\ldots]$, since if it were there would exist a non-zero element a of Q such that $ag = 0$, and this is clearly not the case. Hence the map is injective, and the same is true of the induced map $Q_N \to Q_g$ where Q_N is as defined in chapter 7, section 4. We will identify Q_N with its isomorphic image in Q_g. Then the rings Q_N form an ascending sequence of sub-rings of Q_g whose union is Q_g. We denote the residue field of Q_g by k_g and its maximal ideal by $\mathfrak{m}_g$.

We will also apply the above to the special case when Q is a field F, i.e. F_g and F_N will denote the fields $F(X_1,X_2,\ldots)$ and $F(X_1,X_2,\ldots,X_N)$.

We note further that Q_g, as the union of a set of flat extensions of Q, is itself a flat extension of Q. (See [BAC], Chapter 1, section 2, proposition 2(ii) on p.28.) Since the homomorphism $Q \to Q_g$ is local, it follows further that Q_g is a faithfully flat extension of Q. Hence most of the results of section 4 of chapter 7 remain true if we replace n by g. For convenience we list those contained in 7.41-7.43.

(10.11). *If* M *has finite length, then*

$$l(M) = l(M_g).$$

(10.12). *If* I, J *are ideals of* Q, *then*

i) $(I \cap J)Q_g = IQ_g \cap JQ_g$,

ii) $(I : J)Q_g = IQ_g : J$.

(10.13). *If* J *is an ideal of* Q, *then*

i) *the rings* $(Q/J)_g$ *and* Q_g/JQ_g *are isomorphic,*

ii) *if* J *is prime so is* JQ_g,

iii) *if* M *is a finitely generated* Q-*module and*

$$(0) = N_1 \cap \ldots \cap N_s$$

is an irredundant primary decomposition of the zero sub-module of M, *then*

$$(0) = N_{1g} \cap \ldots \cap N_{sg}$$

is an irredundant primary decomposition of the zero sub-module of M_g.

Further, if N_i *has radical* $\mathbf{p}_i$, N_{ig} *has radical* $\mathbf{p}_{ig} = \mathbf{p}_i Q_g$.

Next in this list of elementary results we record two simple results in the form of a lemma.

LEMMA 10.14. *There exist natural isomorphisms of* Q-*algebras*

$$\alpha_N: (Q_N)_g \to Q_g \text{ and } \beta_N: (Q_g)_N \to Q_g.$$

We consider Q_N as the localisation of $Q[Y_1,\ldots,Y_N]$ at $\mathbf{m}[Y_1,\ldots,Y_N]$, and $(Q_N)_g$ as the localisation of $Q_N[Z_1,Z_2,\ldots]$ at $\mathbf{m}Q_N[Z_1,Z_2,\ldots]$. Then α_N is determined by the facts that it is the identity on Q and the equations

$$\alpha_N(Y_i) = X_i, \quad i = 1,\ldots,N; \quad \alpha_N(Z_j) = X_{N+j}, \quad j = 1,2,\ldots$$

To define β_N we proceed as follows. First define β_N on Q_g by the condition that it is the identity on Q and

$$\beta_N(X_j) = X_{N+j}.$$

Next extend this to a map $Q_g[Y_1,\ldots,Y_N] \to Q_g$ by defining $\beta_N(Y_i) = X_i$, $i = 1,\ldots,N$, and finally extend this to a map $(Q_g)_N \to Q_g$. Then this map is the required isomorphism.

We continue our list of elementary results.

(10.15A). *If* $\mathbf{p}$ *is a prime ideal of* Q *then* $(Q_{\mathbf{p}})_g$ *is isomorphic to the localisation of* Q_g *at* $\mathbf{p}_g$.

(10.15B). $\dim Q_g = \dim Q$.

This is most easily seen as follows. By (10.11) $l(Q_g/\mathfrak{m}_g{}^n) = l(Q/\mathfrak{m}^n)$ for all n. The right-hand side is for large n equal to a polynomial in n whose degree is equal to dimQ, while, again for large n, the left-hand side is equal to a polynomial in n whose degree is $\dim Q_g$. Hence the result.

(10.15C). *For any ideal* I *of* Q

$$\dim(Q_g/I_g) = \dim(Q/I).$$

Immediate from (10.13)i) and (10.15B).

(10.15D). *If* $\mathbf{p}$ *is a prime ideal of* Q, *then*

$$\mathrm{ht}\,\mathbf{p}_g = \mathrm{ht}\,\mathbf{p}.$$

Immediate from (10.15A),(10.15B), since $\mathrm{ht}\,\mathbf{p} = \dim Q_{\mathbf{p}}$.

(10.15E). Q_g *is regular if and only if* Q *is regular.*

If dimQ = d, then Q is regular if and only if $l(\mathfrak{m}/\mathfrak{m}^2) = d$. But

$$l(\mathfrak{m}_g/\mathfrak{m}_g{}^2) = l(\mathfrak{m}/\mathfrak{m}^2)$$

by (10.11). Since $\dim Q_g = \dim Q$, the result follows.

(10.15F). *If* f *is a noether filtration on* Q *with radical* $\mathfrak{m}$, *and* f_g *denotes the filtration on* Q_g *defined by the sequence of ideals* $I_n(f)Q_g$, *then*

$$e(f,Q) = e(f_g,Q_g).$$

Immediate from 8.1, and (10.11).

(10.15G). *Let* I *be an ideal of* Q_N. *Then* $I_g \cap Q_N = I$.

By Lemma 10.14 we need only consider the case N = 0, while by 10.13(i) we can take I = (0). The result then follows from the fact that the map $Q \rightarrow Q_g$ is injective.

The ring Q_g is noetherian; while this is a consequence of a much more general result of Grothendieck ([EGA], Chapter 0, proposition 10.3.1), we give an *ad hoc* proof.

LEMMA 10.16. *If* I *is a finitely generated ideal of* Q_g *then*

$$\cap(I + \mathfrak{m}_g^n) = I,$$

the intersection being taken over all positive integers n.

Let x be an element of $\cap(I+\mathfrak{m}_g^n)$, and let $a_1,\ldots,a_m$ be a set of generators of I. Then we can find an integer N such that $x,a_1,\ldots,a_m$ all belong to Q_N. Further, since $\mathfrak{m}_g^n$ is generated by elements of Q for all n, it follows that Q_N contains sets of generators of $I+\mathfrak{m}_g^n$ for all n. Hence, if $I' = I\cap Q_N$, $x \in (I'+\mathfrak{m}^n)Q_g\cap Q_N = I'+\mathfrak{m}^nQ_N$ for all n. Hence, as Q_N is noetherian, this implies that $x \in I'$ and hence to I. (See Lemma 5.25.)

THEOREM 10.17. Q_g *is noetherian.*

Since the maximal ideal of Q_g is finitely generated, the completion $Q_g\hat{}$ of Q_g is noetherian. Hence if I is any ideal of Q_g, $IQ_g\hat{}$ has a finite basis composed of elements $u_1,\ldots,u_r$ of I. Let I' be the ideal of Q_g generated by $u_1,\ldots,u_r$. Then

$$I \supseteq I' = \cap(I' + \mathfrak{m}_g^n) = I'Q_g\hat{}\cap Q_g \supseteq I$$

proving that I is finitely generated.

DEFINITION. *Let* I *be an ideal of* Q_g, *and let* $I_N = (I\cap Q_N)Q_g$. *Then, for all* N, I_N *is contained in* I, *and the fact that* Q_g *is noetherian implies that* $I_N = I$ *for* N *sufficiently large. The least integer* N *for which this is true will be termed the index of definition of* I *and will be denoted by* def(I). *More generally, if* $\mathbf{I} = (I_1,\ldots,I_s)$ *is a set of ideals,* def(**I**) *is the least integer* N *such that* $(I_j)_N = I_j$ *for* $j = 1,\ldots,s$.

It is clear from 10.13 that, if def(I) = N, then all associated prime ideals $\mathfrak{p}$ of I satisfy def($\mathfrak{p}$) $\le$ N.

2. Prime ideals of Q_g.

The relationship between Q and Q_g is stronger than the statement that Q_g is a faithfully flat extension of Q indicates. To see this we must consider the fibres of the map $Q \to Q_g$. Let $\mathfrak{p}$ be a prime ideal of Q. The fibre of the map $Q \to Q_g$ at $\mathfrak{p}$ is the algebra $k(\mathbf{p})\otimes_Q Q_g$, where $k(\mathbf{p})$ is the field of fractions of $Q/\mathbf{p}$. Then the map $Q \to Q_g$ is regular, where this means, in addition to the map being flat, that for all $\mathbf{p}$ and for all finite field extensions L of $k(\mathbf{p})$, the ring $L\otimes_{k(\mathbf{p})} Q_g$ is a regular ring, i.e., its localisations at all its prime ideals are regular local rings. We now prove this statement.

THEOREM 10.21. *The maps* $Q \to Q_N$ *and* $Q \to Q_g$ *are regular.*

If $\mathbf{p}$ is a prime ideal of Q, then $(Q/\mathbf{p})_N = Q_N/\mathbf{p}Q_N$ and $(Q/\mathbf{p})_g = Q_g/\mathbf{p}Q_g$. Hence it will be sufficient to take Q to be a domain and to take $\mathbf{p} = (0)$. We write K for the field of fractions of Q, so that L is a finite extension of K.

We first prove that the maps $Q \to Q_N$ are regular. $Q_N \otimes_Q L$ is a ring of fractions of $L[X_1,\dots,X_N]$, the denominators belonging to the set S consisting of all polynomials in $X_1, X_2, \dots$ with coefficients in Q, with at least one coefficient a unit of Q. Since $L[X_1,\dots,X_N]$ is regular, so is $L\otimes_Q Q_N$.

Now we turn to Q_g. $L\otimes_Q Q_g$ is a finite module over $K\otimes_Q Q_g$ which, as a ring of fractions of Q_g, is noetherian. Hence $L\otimes_Q Q_g$ is noetherian, and is the union of its sub-rings $L\otimes_Q Q_N$. Let $\mathcal{P}$ be a prime ideal of $L\otimes_Q Q_g$. Then $\mathcal{P}$ is generated by elements of $L\otimes_Q Q_N$ for N large enough. Let $\mathcal{P}_N = \mathcal{P}\cap Q_N$. Then $\mathcal{P} = \mathcal{P}_N Q_g \otimes_Q L$, where Q_N is identified with the sub-ring $1\otimes_Q Q_N$ of $L\otimes_Q Q_g$. It follows that the localisation of $L\otimes_Q Q_g$ at $\mathcal{P}$ is the general extension of the localisation of $L\otimes_Q Q_N$ at $\mathcal{P}_N$. But the latter is regular by the first part of this proof. Hence $(L\otimes_Q Q_g)_{\mathcal{P}}$ is a regular local ring for all choices of $\mathcal{P}$ and consequently $L\otimes_Q Q_g$ is regular. Hence the map $Q \to Q_g$ is regular.

An immediate consequence of the above theorem is that, if Q satisfies one of the conditions

(S_k): $\quad \operatorname{depth}(Q_{\mathbf{p}}) \geq \operatorname{Inf}(k, \operatorname{ht}\mathbf{p})$ for all prime ideals $\mathbf{p}$ of Q,

or,

(R_k): $Q_{\mathbf{p}}$ is a regular local ring for all prime ideals $\mathbf{p}$ of Q such that ht$\mathbf{p} \leq k$,

then Q_g satisfies the corresponding condition. This is a consequence of, for example, [M] 21C Corollary 2 on p.154 and 21E on p.156, since, the fibres of the map $Q \to Q_g$ all being regular, satisfy the conditions (S_k) and (R_k) for all k. This leads to immediate proofs that if Q satisfies any of the conditions: a) it is reduced(i.e., satisfies R_0), or b) it is Cohen-Macaulay, (i.e., satisfies S_k for all k), or c) is normal, (i.e., satisfies R_1 and S_2), then Q_g satisfies the corresponding condition. However we will give *ad hoc* proofs of a) to c) at the end of this chapter.

For our next result we require some new notation. Let N be any positive integer, and write $Q_{[N]}$ for the localisation of the sub-ring of Q_g formed by localising $Q[X_{N+1}, X_{N+2}, \ldots]$ at $\mathbf{m}[X_{N+1}, X_{N+2}, \ldots]$. Then, as in Lemma 10.14, it is clear that $Q_{[N]}$ is isomorphic to Q_g. Now suppose that I is any ideal of Q_g, I_N denotes $I \cap Q_N$, where $N = \mathrm{def}(I)$, so that $I = I_N Q_g$. Write also I_0 for $I \cap Q$. We now have the following simple lemma.

LEMMA 10.22. *If* I,N *are as above, then*

$$I \cap Q_{[N]} = I_0 Q_{[N]}.$$

It is immediate that the left-hand side contains the right-hand side. Now suppose that x is an element of $I \cap Q_{[N]}$. Then we can write x in the form

$$x = f(X_{N+1}, \ldots, X_{N+r})/g(X_{N+1}, \ldots, X_{N+r})$$

where r is a sufficiently large integer, f, g being polynomials over Q such that f belongs to I and g has at least one coefficient a unit of Q. Increasing r, if necessary, we can now write

$$f(X_{N+1}, \ldots, X_{N+r}) = F(X_{N+1}, \ldots, X_{N+r})/G(X_{N+1}, \ldots, X_{N+r}),$$

where F, G are polynomials over Q_N, the coefficients of F being in I_N and at least one coefficient of G being a unit. Then consider the equation

$$f(X_{N+1}, \ldots, X_{N+r})G(X_{N+1}, \ldots, X_{N+r}) = F(X_{N+1}, \ldots, X_{N+r}).$$

The coefficients of the polynomials involved all belong to Q_N, while those of f belong to Q and those of F belong to I_N. Reduce modulo I_N. Then the right-hand side becomes zero, while the image of G remains a non-zero divisor. This implies that the coefficients of f all belong to $I_N \cap Q = I_0 = (0)$, which proves the result.

Our next result is important in the next two chapters. First we need a definition.

DEFINITION *Let* $\mathcal{P}$ *be a prime ideal of* Q_g *and let* $N = \mathrm{def}(\mathcal{P})$. *Let* F *be the field of fractions of* $Q/\mathcal{P}_0$ *and* E_N *be the field of fractions of* $Q_N/\mathcal{P}_N$, *so that* E_N *is an extension of* F *generated by* N *elements. Then we define* $t(\mathcal{P})$ *by the equation*

$$\mathrm{tr.deg.}_F E_N = N - t(\mathcal{P})$$

The last equation is still valid if we only require that $N \geq \mathrm{def}\mathcal{P}$. For suppose that $N' > N = \mathrm{def}\mathcal{P}$ and that $E_{N'}$ is the field of fractions of $Q_{N'}/\mathcal{P}_{N'}$. Then, as $\mathcal{P}_{N'} = \mathcal{P}_N Q_{N'}$, it follows that $E_{N'} = E_N(X_{N+1},\dots,X_{N'})$, i.e. $\mathrm{tr.deg}_F E_{N'} = \mathrm{tr.deg}_F E_N + N' - N$, and hence

$$\mathrm{tr.deg.}_F E_{N'} = N' - t(\mathcal{P}).$$

In the proof of the theorem which follows, we use a technique which will be used again in the course of the two chapters following. It consists of proving a result first for the ring Q_N and then deducing the corresponding theorem for Q_g from it. Note that the above definition of $t(\mathcal{P})$ is an example of this procedure, since it essentially first defines $t(\mathcal{P})$ for a prime ideal of Q_N and then uses this definition to define $t(\mathcal{P})$ for a prime ideal of Q_g. We will, without comment, use $t(\mathcal{P})$ for prime ideals of both Q_N and Q_g below.

THEOREM 10.23. *Let* Q^* *denote either* Q_N *or* Q_g, *let* $\mathcal{P}$ *be a prime ideal of* Q^* *and let* $\mathbf{p} = Q \cap \mathcal{P}$. *Then*

i) $\dim(Q/\mathbf{p}) - \dim(Q^*/\mathcal{P}) \geq t(\mathcal{P}) = \mathrm{ht}\mathcal{P} - \mathrm{ht}\mathbf{p}$,

ii) $R^* = (Q^*)_{\mathbf{p}}/\mathbf{p}(Q^*)_{\mathbf{p}}$ *is regular of dimension* $t = t(\mathcal{P})$.

We first take $Q^* = Q_N$. We apply the altitude inequality to the pair of domains $Q/\mathbf{p}$, $Q_N/\mathcal{P}$. Then the latter is isomorphic to a localisation of a finitely generated extension of $Q/\mathbf{p}$ with field of fractions E_N of transcendence degree $N - t(\mathbf{P})$ over the field of fractions F of $Q/\mathbf{p}$, and the residue field of $Q_N/\mathbf{P}$ clearly has transcendence degree N over the residue field k of $Q/\mathbf{p}$. Hence

$$\dim(Q/p) + N - t(\mathbf{P}) \geq \dim(Q_N/\mathbf{P}) + N$$

which proves the first inequality in i).

We next prove ii). First we note that we can consider R^* as a ring of fractions of $F[X_1,\dots,X_N]$. This implies that R^* is regular. Further, the altitude equality applies in this case and yields

$$N = \dim(R^*) + N - t(\mathbf{P})$$

proving that $\dim R^* = t(\mathbf{P}) = t$.

Now we return to the equality in i). We recall the definition of $Q_{(N)}$ given in Chapter 7, viz., that it is the localisation of $Q[X_1,\dots,X_N]$ at the prime ideal $(\mathbf{m},X_1,\dots,X_N)$. We can consider $(Q_N)_\mathbf{P}$ as a localisation of $(Q_\mathbf{p})_{(N)}$ at a prime ideal $\mathbf{P}^*$, whence $(Q_{(N)})_{\mathbf{P}^*}/\mathbf{p}(Q_{(N)})_{\mathbf{P}^*}$ is isomorphic to R^* and so has dimension t. Now the localisation of $(Q_{(N)})_{\mathbf{P}^*}$ at $\mathbf{p}(Q_{(N)})_{\mathbf{P}^*}$ is simply $(Q_\mathbf{p})_N$, proving that $\mathbf{p}(Q_{(N)})_{\mathbf{P}^*}$ has height $\mathrm{ht}\mathbf{p}$. Hence, by 7.34, $\mathrm{ht}\mathbf{p} + t(\mathbf{P})$ is a quasi-height of the maximal ideal of $((Q_\mathbf{p})_{(N)})_{\mathbf{P}^*}$ which has dimension $\mathrm{ht}\mathbf{P}^* = \mathrm{ht}\mathbf{P}$. Hence

$$\mathrm{ht}\mathbf{P} \geq \mathrm{ht}\mathbf{p} + t(\mathbf{P}).$$

Next consider $Q' = (Q_\mathbf{p})_{(N)}/\mathbf{P}^*$. Again by 7.34, $\mathrm{ht}\mathbf{P}^* + \dim Q'$ is a quasi-height of the maximal ideal of $(Q_\mathbf{p})_{(N)}$, and so is $\dim(Q_\mathbf{p})_{(N)} = \mathrm{ht}\mathbf{p} + N$, i.e.,

$$\mathrm{ht}\mathbf{P} + \dim Q' = \mathrm{ht}\mathbf{P}^* + \dim Q' \leq \mathrm{ht}\mathbf{p} + N. \tag{*}$$

But, as $\mathbf{P}^* \cap Q_\mathbf{p} = \mathbf{p}Q_\mathbf{p}$, Q' is a homomorphic image of $(Q_\mathbf{p})_{(N)}/\mathbf{p}(Q_\mathbf{p})_{(N)}$ which is isomorphic to $F[X_1,\dots,X_N]$ localised at $(X_1,\dots,X_N)$ and so has residue field F. Q' has field of fractions isomorphic to the residue field of $Q^*_\mathbf{p}$, i.e., E_N, which has transcendence degree $N - t(\mathbf{P})$ over F. Since we are dealing with domains finitely generated over a field, it follows that $\dim Q' = N - t(\mathbf{P})$ and hence (*) becomes the inequality

$$\mathrm{ht}\mathbf{P} \leq \mathrm{ht}\mathbf{p} + t(\mathbf{P})$$

which, with our earlier inequality, yields the equation

$$\mathrm{ht}\mathcal{P} = \mathrm{ht}\mathfrak{p} + t(\mathcal{P}).$$

Now we turn to the case $Q^* = Q_g$. For N large enough, $\mathcal{P} = \mathcal{P}_N Q_g$, and i) follows from the above applied to Q_N, $\mathcal{P}_N$ if we note that $\dim(Q_g/\mathcal{P}) = \dim(Q_N/\mathcal{P}_N)$ and $\mathrm{ht}\mathcal{P} = \mathrm{ht}\mathcal{P}_N$, since $Q_g/\mathcal{P} = (Q_N/\mathcal{P}_N)_g$ and $(Q_g)_{\mathcal{P}}$ is the general extension of Q_N localised at $\mathcal{P}_N$. ii) follows since $(Q_g)_{\mathcal{P}}/\mathfrak{p}(Q_g)_{\mathcal{P}}\ R^*$ is the general extension of $(Q_N)_{\mathcal{P}^*}/\mathfrak{p}(Q_N)_{\mathcal{P}^*}$ where $\mathcal{P}^* = \mathcal{P}\cap Q_N$.

3. Valuations on general extensions.

In this section we will be concerned with a more general class of valuations on a noetherian ring than ideal valuations. To be precise, we will consider valuations on a noetherian ring A which take non-negative integer values together with ∞. Thus we have two prime ideals associated with such a valuation v, its centre $\mathfrak{p}$ (consisting of all x such that $v(x) > 0$) and the prime ideal $\mathcal{P}$ on which v takes the value ∞, which we will term the *limit ideal* of v. Now it is clear that we can always extend v to a valuation on the ring $A_{\mathfrak{p}}$ whose centre is therefore the maximal ideal of $A_{\mathfrak{p}}$ and whose limit is the ideal $\mathcal{P}A_{\mathfrak{p}}$. We therefore, for the moment, restrict attention to valuations on a local ring $(Q,\mathfrak{m},k,d)$ whose centre is the maximal ideal $\mathfrak{m}$ of Q. With such a valuation we can associate a function $v(I) = \mathrm{Min}\,(v(x) \mid x\in I)$ on the set of all $\mathfrak{m}$-primary ideals, which, except in the degenerate case where the limit of v is equal to its centre $\mathfrak{m}$, takes finite integer values. We will refer to this function as an $\mathfrak{m}$*-valuation*, and term it *proper* if it takes values other than $0,\infty$. An $\mathfrak{m}$-valuation clearly has the properties that

a) $v(I_1I_2) = v(I_1) + v(I_2)$,

b) if $I_1 \supseteq I_2$, then $v(I_1) \leq v(I_2)$.

We will refer to two $\mathfrak{m}$-valuations v, w as *independent*, if there do not exist non-zero integers m, n such that $mv(I) = nv(I)$ for all I.

LEMMA 10.31. *Let* $v_1,\ldots,v_N$ *be a set of proper* $\mathfrak{m}$*-valuations on a local ring* Q *which are independent in pairs. Then, if real numbers* $a_1,\ldots,a_N$ *satisfy*

$$\Sigma a_i v_i(I) = 0,$$

summed from 1 *to* N, *for all* $\mathfrak{m}$*-primary ideals* I *of* Q, $a_1,\dots,a_N$ *are all zero.*

Let $\mathfrak{P}_i$ be the limit of v_i and let r be the number of distinct prime ideals in the set of prime ideals $\mathfrak{P}_i$. We proceed by induction on r. First suppose that r = 1. Then we can replace Q by $Q/\mathfrak{P}_1$ and therefore assume that the m-valuations $v_1,\dots,v_N$ are derived from valuations on the field of fractions F of Q, which we also denote by $v_1,\dots,v_N$. Now suppose that c is a non-zero element of Q, so that $v_i(c) < \infty$ for all i. Since $v_i(x) > 0$ on $\mathfrak{m}$, we can find an integer n such that $v_i(x) > v_i(c)$ for all x in $\mathfrak{m}^n$ and all i. Then if $I = \mathfrak{m}^n + cQ$, we have, for each i, $v_i(c) = v_i(I)$, and hence

$$\Sigma a_i v_i(c) = 0,$$

and this remains valid if we replace c by any non-zero element of F. But, the hypothesis of pairwise independence of the $\mathfrak{m}$-valuations $v_1,\dots,v_N$ implies that the valuations $v_1,\dots,v_N$ have distinct valuation rings and hence the theorem of independence of valuations implies that the real numbers $a_1,\dots,a_N$ are all zero (see Theorem 9.42).

Now suppose that r > 1. First we choose $\mathfrak{P}$ to be minimal in the set of limit ideals $\mathfrak{P}_1,\dots,\mathfrak{P}_N$ and, renumbering the $\mathfrak{m}$-valuations $v_1,\dots,v_N$ if necessary, we assume that $\mathfrak{P}_i = \mathfrak{P}$ if $1 \le i \le N' < N$, and $\mathfrak{P}_i \neq \mathfrak{P}$ if $i > N'$. Now choose an element c of Q not in $\mathfrak{P}$ but in $\mathfrak{P}_j$ if $j > N'$. Then we can find n such that, if $i \le N'$, $v_i(\mathfrak{m}^n) > v_i(c)$. It follows that

$$v_i(c) = v_i(cQ + \mathfrak{m}^n),\ 1 \le i \le N',$$

while

$$v_j(cQ + \mathfrak{m}^n) = n v_j(\mathfrak{m}),\ \ j > N'.$$

Now clearly these equations still hold if we replace n by n+1, and hence we get

$$\Sigma a_i v_i(c) = 0$$

the sum being from 1 to N'. Now the condition imposed on c is also satisfied if we replace c by ca, where a is any element of Q not in $\mathfrak{P}$. Hence it follows, by the same argument as used in the case r = 1, that $a_i = 0$ if $i \le N'$, and we have now reduced r to

$r - 1$. The result follows by our inductive hypothesis.

We now consider an $\mathfrak{m}_g$-valuation v on Q_g and its restrictions v_0 and v_N to Q and Q_N respectively. First we require a special case of a general result of Abyankhar, the result and the proof given below being taken from [Z-S] Vol. 2, appendix 2. First we require some notation. Let $(Q,\mathfrak{m},k,d)$ be a local domain with field of fractions F and let E be a finitely generated field extension of F. Let v be an integer- valued valuation on E which is ≥0 on Q and has centre $\mathfrak{m}$ on Q. Then the residue field K_v of v is an extension of the field k. We write $t(v/Q)$ for the transcendence degree of K_v over k, and $t(E/F)$ for the transcendence degree of E over F.

LEMMA 10.32. $d + t(E/F) - 1 \geq t(v/Q)$.

Choose elements $z_1,\ldots,z_s$ of the valuation ring O_v of v such that the images of $z_1,\ldots,z_s$ in K_v are algebraically independent over k. Let $\mathfrak{p}$ be the centre of v on the ring $Q[z_1,\ldots,z_s]$ and let Q' denote the localisation of $Q[z_1,\ldots,z_s]$ at $\mathfrak{p}$. Let E' be the field of fractions of Q', so that $t(E'/F) \leq t(E/F)$. Then, applying the altitude inequality to the pair of rings Q, Q'

$$d + t(E/F) \geq d + t(E'/F) \geq \dim Q' + s,$$

which implies that s is bounded above by $d + t(E/F) - 1$, since $\dim Q' > 0$. This implies that K_v has finite transcendence degree over k, and further, we could take $s = t(v/Q)$, yielding the result.

We now turn to the rings Q, Q_g. We suppose that v is a valuation on Q with limit ideal $\mathfrak{P}$. Then we define the general extension v_g of v to Q_g as follows. Let f be an element of $Q[X_1,X_2,\ldots]$. Define $v_g(f)$ to be $\mathrm{Min}(v(a))$, where a runs over the coefficients of f. It is clear that this defines a valuation on $Q[X_1,X_2,\ldots]$ with centre $\mathfrak{m}[X_1,X_2,\ldots]$ and therefore it has a unique extension to Q_g which we also denote by v_g. Clearly, the limit ideal of v_g is $\mathfrak{P}_g$.

LEMMA 10.33. *Let* Q *be a local domain with field of fractions* F *and let* v *be a valuation on* Q *with centre* $\mathfrak{m}$ *and limit ideal* (0). *Then, if* $t(v/Q) = 0$, *and* w *is an extension of* v *to* $F_g \geq 0$ *on* Q_g *also with limit ideal* (0), $w = v_g$.

We commence with an observation. The condition that $w(x) \geq 0$ on Q_g implies that if z is an element of Q_g not in $\mathfrak{m}_g$, so that z^{-1} belongs to Q_g, then $w(z) = 0$. In particular, $w(X_i) = 0$.

Suppose that $w \neq v_g$ on Q_g. Then, as $w = v$ on Q, there exists an integer N such that $w = v_g$ on Q_N but not on Q_{N+1}. Now consider the restriction v_N of v_g to Q_N. The residue field $K(v_N)$ of v_N is $K_v(X_1,\ldots,X_N)$ while the residue field k_N of Q_N is $k(X_1,\ldots,X_N)$. Since K(v) is algebraic over k, it follows that $K(v_N)$ is algebraic over k_N. Hence by replacing Q by Q_N and v by v_N we have reduced the proof to the case $N = 0$.

It now follows that there exists a polynomial $f(X_1) = a_0 + a_1X_1 + \ldots + a_rX_1^r$, with coefficients in F (or even in Q), such that

$$w(f) \neq v_1(f) = \mathrm{Min}\,(v(a_i)).$$

Now $w(f) \geq \mathrm{Min}\,(w(a_iX_1^i)) = \mathrm{Min}\,(v(a_i)) = v_1(f)$. Hence we have $w(f) > \mathrm{Min}\,(v(a_i))$. Choose j so that $v(a_j) = \mathrm{Min}\,(v(a_i))$. Then the above remains true if we replace f by $a_j^{-1}f$. Hence we may assume that all the coefficients a_i of f belong to O_v and that one of the coefficients is 1. Since $w(f) > 0$, the image of X_1 in $K(w_1)$, where w_1 is the restriction of w to $F(X_1)$, is algebraic over K(v) and hence over the image of k in K(v). Now any element of the image of k in K(v) is the image of an element of Q under the map of Q into K(v), i.e., the product of the inclusion map of Q in O_v followed by the natural map of O_v onto K(v). Now let m(Y) be the minimum polynomial of the image of X_1 in $K(w_1)$ over k. This will be the image of a polynomial

$$g(Y) = Y^s + b_1Y^{s-1} + \ldots + b_s$$

with coefficients in Q under the map $Q \rightarrow K(v)$ already indicated. It follows that $g(X_1)$ has zero image in $K(w_1)$ and hence that

$$w(g(X_1)) = w_1(g(X_1)) > 0.$$

But as g(Y) is monic, $g(X_1)$ does not belong to $\mathfrak{m}_g$, and, by our initial observation

$w(g(X_1)) = 0$. Hence the assumption that $w \neq v_g$ leads to a contradiction and our result is proved.

THEOREM 10.34. *Let* w *be a valuation on* Q_g *with centre* $\mathfrak{m}_g$. *Then there exists an integer* N *such that* w *is the general extension of the restriction* w_N of w *to* Q_N.

We find N in two stages. Let $\mathfrak{p}$ be the limit of w. Let $N' = \mathrm{def}(\mathfrak{p})$, so that $\mathfrak{p}$ is generated by elements of $Q_{N'}$. Replacing Q by $Q_{N'}/\mathfrak{p}_{N'}$, where $\mathfrak{p}_{N'} = \mathfrak{p} \cap Q_{N'}$, we can assume that w has centre (0) on Q_g. Next, we can find $N > N'$ such that the field of fractions of Q_N contains elements $z_1, \ldots, z_s$ in O_w whose images in K_w form a transcendence basis of K_w over k_g, the transcendence degree of K_w over k_g being finite by 10.32. Let Q' be the localisation of $Q[z_1, \ldots, z_s]$ at the centre of w. Then $t(w_N/Q') = 0$, and w is a valuation on Q'_g. Hence, by Lemma 10.32, w is the general extension of w_N on the field of fractions of Q'_g, which is the same as that of Q_g.

DEFINITION. *Let* w *be an* $\mathfrak{m}_g$*-valuation on* Q_g *and let* v_N *be its restriction to* Q_N. *Then* $\mathrm{def}(w)$ *is defined to be the least integer such that* w *is the general extension of* v_N.

In fact, the restriction to valuations centre $\mathfrak{m}_g$ is not necessary. Suppose w is a valuation on Q_g whose centre is a prime ideal $\mathfrak{p}$ of Q_g. Then we can find an integer N'' such that $\mathfrak{p}$ is generated by elements of $Q_{N''}$, and we can now replace Q by the localisation of $Q_{N''}$ at the centre $\mathfrak{p}''$ of w, noting that $(Q_g)_{\mathfrak{p}} = ((Q_{N''})_{\mathfrak{p}''})_g$. We can now apply the above argument.

We conclude this section with a different topic, namely that the results of Lemma 7.44 and Theorem 7.45 can be extended with Q_g replacing Q_n. First we define $\mathrm{def}(f)$ for any non-negative noether filtration f on Q_g.

DEFINITION. *If* f *is a non-negative noether filtration on* Q_g, $\mathrm{def}(f)$ *is the least integer* N *such that* f *is generated by elements of* Q_N.

LEMMA 10.35. *If* f *is a non-negative noether filtration on* Q_g *such that* def(f) = N, *then the associated valuations* $v_1,\dots,v_s$ *of* f *satisfy* def(v_i) ≤ N, *and are the general extensions of the valuations associated with the restriction of* f *to* Q_N.

By replacing Q by Q_N, we can assume that f is generated by elements of Q, and hence that f is the general extension of its restriction f_0 to Q, i.e., is defined by the sequence of ideals $I_n(f)Q_g$. If $v_{01},\dots,v_{0s}$ are the associated valuations of f_0, then the argument of Lemma 7.44 needs no essential alteration to prove that the associated valuations of $f = (f_0)_g$ are $v_i = (v_{0i})_g$ for i = 1,...,s.

LEMMA 10.36. i) *Let* (Q,$\mathfrak{m}$,k,d) *be a local ring. Then if* Q *has any of the properties*

a) Q *is Cohen-Macaulay,*

b) Q *is normal,*

c) Q *is analytically unramified,*

then so does Q_g.

ii) *The maximal ideals of* Q *and* Q_g *have the same quasi-heights, implying that* Q_g *is quasi-unmixed if and only if* Q *is quasi-unmixed.*

i) a) It is clear that if there exist elements y_j of Q_g for j = 1,...,i such that

$$\Sigma y_j x_j = 0$$

where $x_1,\dots, x_d$ is a Q-sequence, then for N large enough, $y_1,\dots,y_i$ all belong to Q_N, and hence y_i belongs to $(x_1,\dots,x_{i-1})Q_N$ by 7.45. This proves that $x_1,\dots,x_d$ is a Q_g sequence.

i) b) If Q is normal, so is Q_N for all N, by Theorem 7.44. Now if x = y/z is an element of the field of fractions of Q_g which is integrally dependent on Q_g, then, for N large enough, x belongs to the field of fractions of Q_N and is integrally dependent on Q_N. Hence it is in Q_N and therefore in Q_g.

i) c) Follows almost exactly as in 7.45.

To prove ii), let f be the filtration $f_{\mathfrak{m}}$. Then if the associated valuations of f are $v_1,\dots,v_s$, the set of distinct integers in the set {dim(v_i|Q) + 1: i = 1,...,s) coincides with the set of quasi-heights of Q by Theorem 6.24. The valuations associated with f_g are the general extensions v_{ig} of the valuations v_i and $\dim(v_{ig}|Q_g) = \dim(v_i|Q)$. Hence the set of quasi-heights of Q, Q_g are the same. Since a local ring is quasi-unmixed if and only if all its quasi-heights are equal, it follows that Q_g is quasi-unmixed if and only if Q is quasi-unmixed.

11. GENERAL ELEMENTS

1. **Introduction.**

Let $(Q,\mathfrak{m},k,d)$ be a local ring. In this section we are concerned with the definition of a general element x of an ideal I of Q or, more generally, given a set of ideals $I_1,\dots,I_s$ of Q, of a set of independent general elements $x_1,\dots,x_s$ of the ideals $I_1,\dots,I_s$. The elements $x, x_1,\dots,x_s$ are not elements of Q, but of Q_g or of Q_N for N large. To be precise, x belongs to IQ_g (or IQ_N) and x_j to I_jQ_g (or I_jQ_N).

In the account that follows, for typographical reasons, we will often use alternative notation for certain symbols. We now make this more precise. The elements $X_1, X_2,\dots$ of the countable sequence of indeterminates used in the definition of the ring Q_g will occasionally be written as X(1), X(2),... Similarly, where we have a set of elements indexed by a set of suffixes $i_1, i_2,\dots,i_s$, rather than involve the use of suffix upon suffix, we will use a notation such as $u(i_1,i_2,\dots,i_s)$, but probably not u(i(1),...,i(s). Finally, a sequence of symbols such as $r_1,r_2,\dots,r_s$ may be represented by a single capital letter R, and we will then define R_i to mean the sequence obtained by omitting the ith term of the sequence, i.e., the sequence $r_1,r_2,\dots,r_{i-1},r_{i+1},\dots,r_s$.

In order to give the definition of general elements or sets of independent general elements, we must first consider automorphisms of the ring Q_g over Q. We recall that Q_g is the union of the rings Q_N, where Q_N is the localisation of $Q[X_1,\dots,X_N]$ at the prime ideal $\mathfrak{m}[X_1,\dots,X_N]$. Now suppose that T is an automorphism of the ring Q_N over Q, i.e., $T(x) = x$ if $x \in Q$. Then we have an extension of T to an automorphism of Q_g over Q, if we define $T(X_i) = X_i$ for all $i > N$. These automorphisms of Q_g over Q will form a sub-group $A_N(Q_g)$ of the group of all automorphisms of Q_g over Q. The union of the sub-groups $A_N(Q_g)$ will be denoted by $A(Q_g)$.

DEFINITIONS. *Let* $(Q,\mathfrak{m},k,d)$ *be a local ring and let* $I_1,\dots,I_s$ *be a set of ideals of* Q. *Let* $a(j,1),\dots,a(j,m_j)$ *be a set of generators of* I_j *for* $j = 1,\dots,s$. *Finally let*

$M_j = m_1 + m_2 + \ldots + m_{j-1}$, *in particular* $M_1 = 0$. *Then the set of elements*

$$y_j = \Sigma a(j,k)X(M_j + k), \quad j = 1,\ldots s$$

where the sum is over $k = 1$ *to* m_j, *will be referred to as a standard set of independent general elements of the set of ideals* $I_1,\ldots,I_s$. *If* $s = 1$, *we will refer to* y_1 *as a standard general element of* I_1.

Let $x_1,\ldots,x_s$ *be a set of elements of* Q_g *such that* $x_j \in I_jQ_g$, $j = 1,\ldots,s$. *Then we will term* $(x_1,\ldots,x_s)$ *a set of independent general elements of* $I_1,\ldots,I_s$, *if there exists an automorphism* T *of* Q_g *contained in* $A(Q_g)$ *such that the set* $y_j = T(x_j)$, $j = 1,\ldots,s$, *is a standard set of independent general elements of* $I_1,\ldots,I_s$. *If* $s = 1$, *then we refer to* x_1 *as a general element of* I_1. *If* $I_1 = I_2 = \ldots = I_s = I$, *then we refer to* $x_1,\ldots,x_s$ *as independent general elements of* I.

THEOREM 11.11. *Let* $(Q,\mathfrak{m},k,d)$ *be a local ring and let* $I_1,\ldots,I_s$ *be ideals of* Q. *Let* $x_1,\ldots,x_s$; $x_1',\ldots,x_s'$ *be two sets of independent general elements of the ideals* $I_1,\ldots,I_s$. *Then there exists an automorphism* T *of* Q_g *over* Q *contained in* $A(Q_g)$, *such that*

$$T(x_j) = x_j', \quad j = 1,\ldots,s.$$

From the definition of a set of independent general elements, there is no loss of generality if we assume that both the sets $x_1,\ldots,x_s$; $x_1',\ldots,x_s'$ are standard sets, and, further, we can assume that the sets of generators of $I_1,\ldots,I_s$ involved in the definition of $x_1,\ldots,x_s$ are all minimal sets of generators. Hence we can write

$$x_j = \Sigma a(j,i)X(M_j + i), \quad j = 1,\ldots,s,$$

the sum being from $i = 1$ to $i = m_j$, where $a(j,1),\ldots,a(j,m_j)$ is a minimal basis of I_j. Further we can write

$$x_k' = \Sigma b(k,l)X(N_k + l), \quad k = 1,\ldots, s,$$

the sum being from $l = 1$ to n_k, where $N_k = n_1 + \ldots + n_{k-1}$, and $b(k,1),\ldots,b(k,n_k)$ is a basis of I_k, not necessarily minimal. Clearly $n_j \geq m_j$ for all j.

We now construct T as a product T_2T_1, where T_1,T_2 are elements of $A(Q_g)$. T_1

is defined by a permutation of $X(1),\ldots,X(N_{s+1})$ in which

$$T_1(X(M_j + i)) = X(N_j + i)$$

for $j = 1,\ldots, s$ and $i = 1,\ldots,m_j$, the permutation being completed arbitrarily.Then T_2 has to satisfy the requirement that

$$T_2(\Sigma a(j,k)X(N_j + k)) = (\Sigma b(j,i)X(N_j + i)),\ \ j = 1,\ldots,s$$

the sum on the left being from $k = 1$ to m_j and that on the right from $i = 1$ to n_j.

Now furthermore we can write

$$b(j,i) = \Sigma a(j,k)c_j(k,i),\ \ j = 1,\ldots,s;\ \ i = 1,\ldots,n_j$$

summed from $k = 1$ to m_j. We can now write the condition that we require as

$$\Sigma a(j,k)(T_2(X(N_j + k)) - \Sigma c_j(k,i)X(N_j + i)) = 0,\ \ j = 1,\ldots,s$$

and hence it will suffice if we can choose T_2 so that

$$T_2(X(N_j + k)) = \Sigma c_j(k,i)X(N_j + i),\ \ j = 1,\ldots,s;\ \ k = 1,\ldots,m_j.$$

The matrix $\mathbf{C}(j) = (c_j(k,i))$ has m_j rows and n_j columns. As $b(j,1),\ldots,b(j,n_j)$ form a basis of I_j, it follows that there is a minimal basis of I_j consisting of m_j of these elements. Suppose that these elements are $b(j,i(1)),\ldots,b(j,i(m_j))$. Then the minor formed by columns $i(1),\ldots,i(m_j)$ of $\mathbf{C}(j)$ must have determinant a unit of Q (since we could express the elements $a(j,k)$ in the form $\Sigma d_j(k,r)b(j,i(r))$, implying that the matrix formed by the columns of this minor has an inverse). It follows that we can add rows taken from the $n_j x n_j$ unit matrix to obtain an $n_j x n_j$ matrix $(c_j(k,i))$, with determinant a unit of Q, whose first m_j rows are those of $\mathbf{C}(j)$. We now define

$$T_2(X(N_j + k)) = \Sigma c_j(k,i)X(N_j + i)$$

for $j = 1,\ldots,s$ and $k = 1,\ldots,N_j$, the sum being from $i = 1$ to n_j. If $r > N_{s+1}$, we define $T_2(X(r)) = X(r)$.

COROLLARY i). *If $x_1,\ldots,x_s$ is a set of independent general elements of a set of ideals $I_1,\ldots,I_s$ of Q, and if $i \to \pi(i)$ is a permutation of 1 to s, then the set $x_{\pi(i)}$, $i = 1,\ldots,s$, is a set of independent general elements of the set of ideals $I_{\pi(1)},\ldots,I_{\pi(s)}$.*

We can take $x_1,\ldots,x_s$ to be a standard set of general elements, and we can now take T to be defined by a suitable permutation of $1,\ldots,N$ for N large enough.

COROLLARY ii). *To within an isomorphism as a Q-algebra,* $L(I) = Q_g/(x_1,\ldots,x_s)$ *depends only on the set of ideals* $I_1,\ldots,I_s$ *and not on the choice of the particular set of independent general elements.*

COROLLARY iii). *The ideal* $(x_1,\ldots,x_s)\cap Q$ of Q *depends only on the set of ideals* $I_1,\ldots,I_s$.

The proofs of Corollaries ii) and iii) are both immediate.

2. The ideal generated by a set of independent general elements.

In this section we will be concerned with the ideal generated by a set of independent general elements $(x_1,\ldots,x_s)$ of a set of ideals $(I_1,\ldots,I_s)$ of a local ring $(Q,\mathfrak{m},k,d)$, our particular concern being with its minimal associated prime ideals.

In the following theorem, we again write $\mathfrak{p}$ for $\mathfrak{P}\cap Q$ for typographical reasons.

THEOREM 11.21. *Let* $(Q,\mathfrak{m},k,d)$ *be a local ring,* $I_1,\ldots,I_s$ *be a set of ideals of* Q, *and let* $x_1,\ldots,x_s$ *be a standard independent set of general elements of* $I_1,\ldots,I_s$. *Let* Q^* *denote either* Q_N, *where* N *is large enough for* $x_1,\ldots,x_s$ *to belong to* Q_N, *or* Q_g. *Let* $\mathfrak{P}$ *be a minimal prime ideal of the ideal* $(x_1,\ldots,x_s)Q^*$ *and let* $\mathfrak{p}$ *be the ideal* $Q\cap\mathfrak{P}$. *Then*

i) $\mathfrak{p}Q_g + (x_1,\ldots,x_s)Q^* = \mathfrak{P}\cap\mathfrak{a}$,

where $\mathfrak{a}$ *is an ideal of* Q^* *which contains an element of* Q *not in* $\mathfrak{p}$,

ii) $R^* = (Q^*)_{\mathfrak{P}}/\mathfrak{p}(Q^*)_{\mathfrak{P}}$ *is regular of dimension* $t = t(\mathfrak{P})$, *and* $I_1,\ldots,I_s$ *can be so numbered that the images of* $x_1,\ldots,x_t$ *generate the maximal ideal of* $(Q^*)_{\mathfrak{P}}/\mathfrak{p}(Q^*)_{\mathfrak{P}}$ *and* $I_{t+1},\ldots,I_s$ *are contained in* $\mathfrak{p}$,

iii) *different prime ideals of* Q^* *minimal over* $(x_1,\ldots,x_s)Q^*$ *have different intersections with* Q,

iv) *if* $\mathfrak{P}_1,\ldots,\mathfrak{P}_n$ *are the prime ideals of* Q^* *minimal over* $(x_1,\ldots,x_s)Q^*$ *and* $\mathfrak{p} = \mathfrak{P}_i\cap Q$, *then the set* $\mathfrak{p}_1,\ldots,\mathfrak{p}_n$ *contains all the prime ideals of Q minimal over the ideal* $X(I) = (x_1,\ldots,x_s)Q^*\cap Q$, *and* $\mathfrak{p}_i \supseteq X(I)$ *for all* i.

Finally, if $Q^* = Q_g$, *then* i)-iv) *are true if* $(x_1,\dots,x_s)$ *is only required to be an independent set of general elements of* $I_1,\dots,I_s$.

We first consider the case where $Q^* = Q_N$, so that the elements $x_1,\dots,x_s$ are in standard form.

i) By hypothesis, $x_1,\dots,x_s$ are linear forms in $X_1,\dots,X_N$ with coefficients in Q and so belong to the sub-ring $S = Q[X_1,\dots,X_N]$ of Q^*. Let $P = \mathfrak{P}\cap S$, so that $P\cap Q = \mathfrak{p}$, and $S_P = (Q^*)_{\mathfrak{P}}$. Write Q' for $Q/\mathfrak{p}$, S' for $S/\mathfrak{p}S = Q'[X_1,\dots,X_N]$, $R^* = (Q^*)_{\mathfrak{P}}/\mathfrak{p}(Q^*)_{\mathfrak{P}} = S_P/\mathfrak{p}S_P$ and P' for $P/\mathfrak{p}S$. Then by Theorem 10.23, R^* is a regular local ring of dimension $t = t(\mathfrak{P})$. Further, if F denotes the field of fractions of Q' and E denotes the fields of fractions of S/P(or $Q^*/\mathfrak{P}$), then the definition of $t(\mathfrak{P})$ and the remarks immediately following that definition imply that

$$\mathrm{tr.deg}_F E = N - t(\mathfrak{P}).$$

Now P' is a prime ideal minimal over the ideal of S' generated by the images $x'_1,\dots,x'_s$ of $x_1,\dots,x_s$ under the homomorphism $S \to S'$. Renumber $x_1,\dots,x_s$ so that $x'_1,\dots,x'_h$ are non-zero and $x'_{h+1},\dots,x'_s$ are zero, i.e., $x_{h+1},\dots,x_s$ belong to $\mathfrak{p}S$. Now form the ring of fractions $D = F[X_1,\dots,X_N]$ of the domain S' with respect to the set of non-zero elements of Q' and consider S" as a sub-ring of D. Then $x'_1,\dots,x'_h$ are linearly independent linear forms over Q', since no indeterminate X_i occurs in more than one of them by the definition of standard independent sets, and hence the ideal P" they generate in D is a prime ideal of height h contained in, and so equal to, P'D. It follows that, in S',

$$(x'_1,\dots,x'_h)S' = P'\cap\mathfrak{a}',$$

where $\mathfrak{a}'$ is an ideal of S' with non-zero intersection with Q'. Hence the maximal ideal of $S'_{P'} = R^*$ is generated by the images of $x'_1,\dots,x'_h$ and these form a minimal basis of this ideal. Hence, since R^* is already known to be regular of dimension t, we must have t = h. Further, lifting to S we have

$$\mathfrak{p}S + (x_1,\dots,x_s)S = P\cap\mathfrak{a}''$$

where $\mathfrak{a}''$ is an ideal of S containing an element of Q not in $\mathfrak{p}$. Finally, localising at $\mathfrak{m}S$, we obtain

$$\mathfrak{p}Q_g + (x_1,\dots,x_s)Q^* = \mathfrak{P}\cap\mathfrak{a},$$

where $\mathfrak{a}$ is an ideal of Q^* which contains an element of Q not in $\mathfrak{p}$, and this is the statement of i) for $Q^* = Q_N$.

ii) We have already proved the first statement of ii) and that $h = t$, and the renumbering given in the proof of i) does have the property that $x'_1,\dots,x'_t$ form a basis of the maximal ideal of Q', and that $x_{t+1},\dots,x_s$ belong to $\mathfrak{p}S$ and hence to $\mathfrak{p}Q^*$. Thus the linear forms $x_{t+1},\dots,x_s$ have zero images in $Q^*/\mathfrak{p}Q^*$ and therefore their coefficients all belong to $\mathfrak{p}$. But these coefficients generate I_j. Hence $\mathfrak{p} \supseteq I_j$, if $j > t$.

iii) P' is the only prime of S' minimal over $(x'_1,\dots,x'_t)$ meeting Q' in zero. Hence P is the only prime ideal of S minimal over $(x_1,\dots,x_s)$ meeting Q in $\mathfrak{p}$. Hence, finally $\mathfrak{P}$ is the only prime ideal of Q^* minimal over $(x_1,\dots,x_s)Q^*$ meeting Q in $\mathfrak{p}$. This applied to each prime ideal of Q^* minimal over $(x_1,\dots,x_s)Q^*$ yields iii) for $Q^* = Q_N$.

iv) It is clear that $\mathrm{rad}(X(I)) = \mathrm{rad}((x_1,\dots,x_s)Q^*)\cap Q$, from which it follows that

$$\mathrm{rad}(X(I)) = \mathfrak{p}_1\cap\dots\cap\mathfrak{p}_n,$$

which is equivalent to the two statements in iv).

Now we turn to the case $Q^* = Q_g$. We note that, if $\mathfrak{P}$ is any prime ideal minimal over $(x_1,\dots,x_s)Q_g$, then $\mathrm{def}\,\mathfrak{P} \le N$, if $x_1,\dots,x_s$ belong to Q_N, by the remark following the definition of $\mathrm{def}(I)$. Hence $\mathfrak{P} = \mathfrak{P}_N Q_g$ and since i) is proved for $Q^* = Q_N$ and $\mathfrak{P} = \mathfrak{P}_N$, i) now follows for $Q^* = Q_g$ by multiplying through by Q_g.

Now we consider ii). Define R to be $(Q_g)_{\mathfrak{P}}/\mathfrak{p}(Q_g)_{\mathfrak{P}}$ and let R_N denote the result of replacing Q_g by Q_N and $\mathfrak{P}$ by $\mathfrak{P}_N$ in this expression. Then R is the general extension of R_N, and hence ii) for $Q^* = Q_g$ follows from the case $Q^* = Q_N$ already proved.

Next iii) for $Q^* = Q_g$ follows from iii) for $Q^* = Q_N$ if we note that the prime ideals of Q_g minimal over $(x_1,\dots,x_s)Q_g$ are of the form $\mathfrak{P}_N Q_g$, where $\mathfrak{P}_N$ ranges over the prime ideals of Q_N minimal over $(x_1,\dots,x_s)Q_N$.

iv) follows in the same way as for $Q^* = Q_N$.

The last remark follows since any independent set of general elements of I is the image of $x_1,\dots,x_s$ under a suitable automorphism T of Q_g which leaves the elements of Q fixed.

COROLLARY i). *Let* $I_1,\dots,I_s$ *be ideals having the same radical* J. *Let* $\mathfrak{P}$ *be a minimal prime ideal of* $(x_1,\dots,x_s)Q_g$ *and let* $\mathfrak{p} = \mathfrak{P} \cap Q$. *Then:*

a) *if* $s > t(\mathfrak{P})$, $\mathfrak{P} = \mathfrak{p}_g$ *and* $\mathfrak{p}$ *is a minimal prime ideal of* J;

b) *if* $s = t(\mathfrak{P})$, $\mathfrak{p}$ *is a minimal prime ideal of* Q.

a) Let $\mathfrak{P}$ be as above. Then $\mathfrak{p}_g$ contains x_j for some j, and so contains I_j, and hence J and so all the elements x_i. Thus, as $\mathfrak{P}$ is minimal over $(x_1,\dots,x_s)$, $\mathfrak{P} = \mathfrak{p}_g$. Now suppose $\mathfrak{p} \supseteq \mathfrak{p}' \supseteq J$. Then, $\mathfrak{p}'$ contains all the ideals I_j and hence $\mathfrak{p}'_g$ contains all the elements x_j. Since $\mathfrak{p}_g$ is minimal over $(x_1,\dots,x_s)Q_g$ it follows that $\mathfrak{p}'_g = \mathfrak{p}_g$, and hence that $\mathfrak{p} = \mathfrak{p}'$. Hence $\mathfrak{p}$ is minimal over J.

b) If $s = t(\mathfrak{P})$, then $(Q_g)_{\mathfrak{P}}/\mathfrak{p}(Q_g)_{\mathfrak{P}}$ is regular of dimension s, implying that $\mathfrak{P}/\mathfrak{p}_g$ has height s. But $\mathfrak{P}$ is minimal over an ideal generated by s elements and hence $\mathrm{ht}\mathfrak{P} \leq s$. Hence $\mathfrak{p}_g$ has height 0, i.e. is minimal, implying that $\mathfrak{p}$ is a minimal prime ideal of Q.

COROLLARY ii). *Let* $(Q,\mathfrak{m},k,d)$ *be a local ring,* $I_1,\dots,I_d$ *be* $\mathfrak{m}$*-primary ideals of* Q *and let* $x_1,\dots,x_d$ *be independent elements of* $I_1,\dots,I_d$. *Then the ideal* $\mathfrak{X} = x_1Q_g + \dots + x_dQ_g$ *is* $\mathfrak{m}_g$*-primary.*

By Corollary i), if $\mathfrak{P}$ is any minimal prime ideal over $\mathfrak{X}$ such that $t(\mathfrak{P}) < d$, then $\mathfrak{P} = \mathfrak{m}_g$ and hence $\mathfrak{P}$ is the only prime ideal minimal over $\mathfrak{X}$. Hence $\mathfrak{X}$ is $\mathfrak{m}$-primary. On the other hand, if $t(\mathfrak{P}) = d$, then by 10.23 i), $\mathrm{ht}\mathfrak{P} = \mathrm{ht}\mathfrak{p} + d \geq d$ and this implies $\mathfrak{P} = \mathfrak{m}_g$.

We are going to apply the above to two particular situations.

The first is when the ideals $I_1,\dots,I_s$ are all equal. To be precise, we consider a single ideal I and a sequence of independent general elements $x_1, x_2,\dots$ of I. We are now going to relate the sequence of ideals $(x_1,\dots,x_s)\cap Q$ to the analytic spread of I.

We begin by recalling the original definition of the analytic spread of an ideal I of a local ring $(Q,\mathfrak{m},k,d)$ given in Chapter 6. Let I have basis $(a_1,\dots,a_m)$ and let G(I) denote the ring $Q[ta_1,\dots,ta_m,u]$ and $F(I,\mathfrak{m})$ denote the ring $G(I)/\mathfrak{m}G(I+uG(I))$. Then $F(I,\mathfrak{m})$ is a graded ring over k generated by elements of degree 1. The analytic spread of I is then the spread of the ring $F(I,\mathfrak{m})$. Now, in Chapter 1, the spread of F(I,m) was defined to be the least integer a such that $n^{-a}d(n) \to 0$ as $n \to \infty$, where d(n) is the dimension over k of $F(I,\mathfrak{m})_n$. For our present purposes it is preferable to state the definition as follows. As shown in Northcott and Rees[1953a], d(n) is equal to a polynomial in n for n large, and then the analytic spread a is defined by the statement that this polynomial has degree a-1.

Now suppose we replace I by I_g. Then the ring $F(I_g,\mathfrak{m}_g)$ is simply $F(I,\mathfrak{m})\otimes_k k_g$, from which it follows that I and I_g have the same analytic spread.

We now find it convenient to give yet another definition of analytic spread. Let $\mathfrak{p}$ be the maximal graded ideal of $F(I,\mathfrak{m})$, L(I) denote the local ring $F(I,\mathfrak{m})_{\mathfrak{p}}$ and let $\mathfrak{n}$ be its maximal ideal. Then it is easy to see that $F(\mathfrak{n},\mathfrak{n})$ is isomorphic to $F(I,\mathfrak{m})$ and hence that for large r, $l(\mathfrak{n}^r/\mathfrak{n}^{r+1})$ is equal to a polynomial of degree a-1 in r. This implies that L(I) has dimension a. Hence we can define the analytic spread s(I) of I to be equal to dimL(I). Note that $L(I_g)$ is isomorphic to $L(I)_g$.

Next we recall that $J = (y_1,\dots,y_r)$ is a reduction of I if

$$I^nJ = I^{n+1}$$

for some n. By Nakayama's lemma, this is equivalent to

$$I^nJ + I^{n+1}\mathfrak{m} = I^{n+1}$$

and hence to the condition that the images of $y_1t,\dots,y_rt$ in $F(I,\mathfrak{m})$ should generate a $\mathfrak{p}$-primary ideal. This is equivalent to the statement that the images of $y_1t,\dots,y_rt$ in L(I) generate an $\mathfrak{n}$-primary ideal.

For our first result we require two lemmas.

LEMMA 11.22. *Let* $(Q,\mathfrak{m},k,d)$ *be a local ring and let* I *be an ideal of* Q *whose analytic spread is* a. *Let* $x_1,\dots,x_a$ *be independent general elements of* I. *Then the ideal* $(x_1,\dots,x_a)$ *is a reduction of* I_g.

It is easy to see that the images of $x_1t,\dots,x_at$ in L(I) form a set of independent general elements of $\mathfrak{n}$ and further dimL(I) = a. Hence by Corollary ii) above, $x_1,\dots,x_a$

generate an $\mathfrak{n}L(I_g)$-primary ideal of $L(I_g)$. But this implies that $(x_1,\dots,x_a)$ is a reduction of I_g.

LEMMA 11.23. *Let* $(Q,\mathfrak{m},k,d)$ *be a local ring,* I *be an ideal of* Q, *and let* $x_1,\dots,x_a$ *be independent general elements of* I, *which belong to* Q_N. *Then the elements* $z_i = x_i - X_{N+i}x_a$, $i = 1,\dots,a-1$, *are independent general elements of* I.

Let I have a basis $(a_1,\dots,a_m)$. Then by applying a suitable automorphism, we can take x_i of the form $\Sigma X_{(i-1)m+j}a_j$, the sum being from 1 to m. We now consider the automorphism T of Q_g over Q defined by

$$T(X_{(i-1)m+j}) = X_{(i-1)m+j} - X_{N+i}X_{(a-1)m+j}, \quad i = 1,\dots,a-1;\ j = 1,\dots,m,$$

$$T(X_n) = X_n \quad \text{if } n > (a-1)m.$$

Then $T(x_i) = x_i - X_{N+i}x_a$.

THEOREM 11.24. *Let* I *be an ideal of* $(Q,\mathfrak{m},k,d)$ *of analytic spread* a. *Let* $X_s(I)$ *denote the ideal* $(x_1,\dots,x_s)Q_g\cap Q$, *where* $x_1,\dots,x_s$ *is an independent set of general elements of* I. *Then, if* $s < a$, $X_{s(I)}$ *is contained in the ideal* $(0 : I^n)$ *for n sufficiently large. If* $s \geq a$ *then* $X_s(I)$ *has the same radical as* I *and is contained in* I. *If* $s \geq m$, *where* m *is the minimal number of generators of* I, *then* $X_s(I) = I$.

We begin by imposing the condition on I that it contain a non-zero divisor of Q. It follows that the elements x_i are all non-zero divisors. For the case $s < a$ it is sufficient to prove the case where $s = a - 1$. Under the given restriction we have to show that $(x_1,\dots,x_{a-1})\cap Q$ is zero. Now consider the ring $L = Q[x_1/x_a,\dots,x_{a-1}/x_a]$. This is a homomorphic image of $Q[Y_1,\dots,Y_{a-1}]$, the kernel containing the elements $x_i - x_aY_i$. Now, if $f(Y_1,\dots,Y_{a-1})$ is a polynomial in $Y_1,\dots,Y_{a-1}$ over Q_g such that

$$f(x_1/x_a,\dots,x_{a-1}/x_a) = 0,$$

then $f(Y_1,\dots,Y_{a-1}) = F(Y_1,\dots,Y_{a-1},1)$, where $F(Y_1,\dots,Y_a)$ is a homogeneous polynomial over Q_g such that $F(x_1,\dots,x_a) = 0$. But as $x_1,\dots,x_a$ is a reduction of I and I has analytic spread a, it follows that $x_1,\dots,x_a$ are analytically independent, this meaning that

$F(Y_1,\dots,Y_a)$ has all its coefficients in $\mathfrak{m}_g$. This implies that $L/\mathfrak{m}L$ is isomorphic to $k_g[Y_1,\dots,Y_{a-1}]$ and hence we have a homomorphism of $(Q_g)_{a-1}$ onto the localisation of L at $\mathfrak{m}L$, and the kernel of this homomorphism meets Q_g in (0). But this kernel contains the elements $z_i = x_i - Y_i x_a$, $i = 1,\dots,a-1$, and these are a general set of elements of I by Lemma 11.24. This completes the proof in the case $s < a$ except for the condition that I contains a non-zero divisor. The general case now follows if we consider the ring Q/J, where J is the union of the ideals $(0 : I^n)$ for all n. For the image $I' = I + J/J$ of I in Q' has zero annihilator, and hence contains a non-zero divisor. Now suppose that $s < a$. Then the images $x_1',\dots,x_s'$ of $x_1,\dots,x_s$ in Q'_g are independent general elements of I'. Hence an element of the ideal $X_s(I')$ is zero, and therefore if $z \in X_s(I)$, its image in Q' is zero, i.e., $z \in J$.

If $s \geq a$ then $(x_1,\dots,x_s)Q_g$ has the same radical as IQ_g since it is a reduction of IQ_g. Hence it contains a power, say $I_g{}^k$, of I_g and so its intersection with Q contains I^k (and is also contained in I). Finally, since $x_1,\dots,x_s$ are linearly independent modulo $\mathfrak{m}IQ_g$ if $s \leq \dim(I/I\mathfrak{m})$, it follows that, if $s \geq m$, $x_1,\dots,x_s$ generate I_g, implying that $X_s(I)$ contains I, and so is equal to it.

We now turn to the second case we wish to consider. We first require a definition.

DEFINITION. *A set of ideals* $I_1,\dots,I_s$ *of* $(Q,\mathfrak{m},k,d)$ *is said to be independent if some independent set of general elements* $x_1,\dots,x_s$ *of* $I_1,\dots,I_s$ *is a sub-set of a set of parameters of* Q_g.

We now draw some conclusions from this restriction on $I_1,\dots,I_s$, or rather, on the set of elements $x_1,\dots,x_s$. Note first that the condition holds for one set of independent general elements only if it holds for all such sets, and so is a condition on the set of ideals $I_1,\dots,I_s$.

Now suppose that $(a_1,\dots,a_d)$ is a set of parameters of a local ring $(Q,\mathfrak{m},k,d)$ and that $s \leq d$. Then the height of any minimal prime ideal $\mathfrak{p}$ of $(a_1,\dots,a_s)$ is at most s, while the dimension of the local ring $Q/\mathbf{p}$ is at most d-s. We now give a definition.

DEFINITION. *If* $(Q,\mathbf{m},k,d)$ *is a local ring, then a prime ideal* $\mathbf{p}$ *of* Q *is said to be a good prime ideal if*

$$\mathrm{ht}\mathbf{p} + \dim(Q/\mathbf{p}) = d.$$

We are concerned below with minimal prime ideals which are good in the above sense and which are associated with an ideal J generated by a sub-set $(x_1,\ldots,x_s)$ of a set of parameters. It is a fact that at least one prime ideal associated with J is good. We do not need this result in this chapter, except insofar that it indicates that Theorem 11.25 below is not vacuous, and therefore the proof will be delayed to the next chapter. We will be particularly concerned with the case when the set $(x_1,\ldots,x_s)$ is the set of general elements of a set of ideals $\mathbf{I} = (I_1,\ldots,I_s)$. Note this is equivalent to the statement that $\mathbf{I}$ is an independent set of ideals.

THEOREM 11.25. *Let* $(Q,\mathbf{m},k,d)$ *be a local ring, and* $I_1,\ldots,I_s$ *be an independent set of ideals of* Q. *Let* $\mathbf{x} = (x_1,\ldots,x_s)$ *be a standard independent set of general elements of* $I_1,\ldots,I_s$, *and* Q^* *denote either* Q_g *or* Q_N *where* $N \geq \mathrm{def}(\mathbf{x})$. *Finally, let* $\mathbf{P}$ *be a good prime ideal minimal over the ideal* $(x_1,\ldots,x_s)Q^*$ *and* $\mathbf{p} = \mathbf{P}\cap Q$. *Then*

i) $\mathbf{p}$ *is a good prime ideal of* Q,

ii) *if* $t = t(\mathbf{P})$, *then* $\mathbf{P}$ *contains exactly* $s - t$ *of the ideals* $I_1,\ldots,I_s$. *If these ideals are numbered so that* $\mathbf{P}$ *contains* $I_1,\ldots,I_{s-t}$, *then* $\mathbf{p}Q^*$ *is a minimal prime ideal of* $(x_1,\ldots,x_{s-t})$, *of height* $s-t$,

iii) $\mathbf{p}Q^*$ *is the only minimal prime ideal of* $(x_1,\ldots,x_{s-t})$ *contained in* $\mathbf{P}$.

Finally, if $Q^* = Q_g$, *then the restriction to standard form can be dropped.*

We are assuming that the set of elements $\mathbf{x}$ is in standard form. To be precise, we assume that $a(i,j)$, $j = 1,\ldots,m_i$, is a minimal basis of I_i, and that, for $i = 1,\ldots,s$,

$$x_i = \Sigma a(i,j)X(M_i + j),$$

summed from $j = 1$ to m_i, where as before, $M_i = m_1+\ldots+m_{i-1}$.

i) We first observe that, by Theorem 10.23 i), we have

$$\dim(Q/\mathbf{p}) + \mathrm{ht}\mathbf{p} \geq \dim(Q_g/\mathfrak{P}) + \mathrm{ht}\mathfrak{P}.$$

Now the right-hand side is equal to d since $\mathfrak{P}$ is good, while the left-hand side is at most d (this being the case for all prime ideals of Q). Hence the left-hand side is equal to d and $\mathbf{p}$ is good.

ii) We now have equations

$$\dim(Q/\mathbf{p}) - \dim(Q_g/\mathfrak{P}) = t = \mathrm{ht}\mathfrak{P} - \mathrm{ht}\mathbf{p}.$$

But $\mathfrak{P}$ has height s and dimension d - s in view of the assumption it is good, and $\mathbf{p}$ has height s - t and dimension d - s + t. We further observe that, by Theorem 11.21, $\mathbf{p}$ must contain s - t of the ideals $I_1,\dots,I_s$, which we will take to be $I_1,\dots,I_{s-t}$, and hence $\mathbf{p}Q^*$ contains the ideal $(x_1,\dots,x_{s-t})$. Now $Q^*/\mathbf{p}Q^*$ has dimension d - s + t, and contains a minimal prime ideal $\mathbf{p}'$ of $(x_1,\dots,x_{s-t})$ which has dimension at most d - s + t. It follows that we must have $\mathbf{p}Q^* = \mathbf{p}'$ and $\mathbf{p}Q^*$ is a minimal prime ideal of $(x_1,\dots,x_{s-t})$.

iii) Now let $N' = M_{s-t+1}$, so that $\mathrm{def}(x_1,\dots,x_{s-t}) = N'$. It follows that any minimal prime ideal $\mathbf{p}'$ of $(x_1,\dots, x_{s-t})$ has a basis in $Q_{N'}$ (i.e., $\mathrm{def}\mathbf{p}' \leq N'$). Note that $N \geq N'$, so that Q^* can be written as either $(Q_{N'})_{N-N'}$ or as $(Q_{N'})_g$. Hence, to prove iii), it will be sufficient to prove that $\mathfrak{P}\cap Q_{N'} = \mathbf{p}Q_{N'}$, since this would imply that $\mathbf{p}Q^* \supseteq (\mathbf{p}'\cap Q_{N'})Q^* = \mathbf{p}'$. We now consider the rings $L = Q_{N'}/\mathbf{p}Q_{N'}$ and $L^* = Q^*/\mathbf{p}Q^*$, and the ideal of L^* generated by $x'_{s-t+1},\dots,x'_s$, where x'_i denotes the image of x_i in L^* (note that $x'_i = 0$ if $i \leq s - t$). Now $\mathfrak{P}$ does not contain any of the ideals $I_{s-t+1},\dots,I_s$ and hence $\mathfrak{P}' = \mathfrak{P}/\mathbf{p}Q^*$ does not contain any of the ideals $I'_i = (I_i + \mathbf{p})/\mathbf{p}$ for $i > s - t$. Hence $x'_{s-t+1},\dots,x'_s$ is an independent set of general elements of the ideals $I'_{s-t+1}L,\dots,I'_sL$, since the indeterminates X_j involved in x_i', $i > s - t$, have $j > N'$. But $\mathfrak{P}'$ is a good prime ideal of L^* minimal over $(x'_{s-t+1},\dots,x'_s)Q^*$ since

$$\mathrm{ht}\mathfrak{P}' + \dim L^*/\mathfrak{P}' = t + d - s = \dim L^*.$$

Further, $\mathfrak{P}'$ does not contain any of the ideals $I'_{s-t+1}L,\dots,I'_sL$. Hence, if we apply ii) to the rings L, L^*, and the ideals $I'_{s-t+1}L,\dots,I'_sL$, $\mathfrak{P}'$, we see that $\mathrm{ht}(\mathfrak{P}'\cap L) = 0$, i.e., $\mathfrak{P}'\cap L' = (0)$ by ii) and hence, as required, $\mathfrak{P}\cap Q_{N'} = \mathbf{p}Q_{N'}$. The last remark follows in the same way that the similar remark does in Theorem 11.21.

We conclude this section by returning to a question left open in Theorem 11.21 iv), namely, whether the intersections $\mathfrak{p}_i$ of the prime ideals $\mathfrak{P}_i$ minimal over $(x_1,\ldots,x_s)$ with Q are necessarily minimal over $X(I) = (x_1,\ldots,x_s)\cap Q$. We show that this need not be the case, even if s = 1, with Q a domain, as the following example shows. Let $(Q,\mathfrak{m},k,d)$ be a Cohen-Macaulay local domain with $d \geq 3$. Let $I = I_1$ be the ideal (ab_1,ab_2), where $a \neq 0$ and (a,b_1,b_2) is a Q-sequence contained in $\mathfrak{m}$, so that $b_1Q:b_2 = b_1Q$. Then if $x_1 = ab_1X_1 + ab_2X_2$, so that x_1 is a general element of I, the minimal primes over x_1Q_g are then those minimal over either aQ_g or $(b_1X_1 + b_2X_2)Q_g$. Those minimal over the former meet Q in the prime ideals minimal over aQ and so are non-zero. On the other hand $\mathfrak{P}' = (b_1X_1 + b_2X_2)Q[X_1,X_2]$ is the kernel of the map of $Q[X_1,X_2]$ onto the domain $Q[tb_1,tb_2]$ in which X_1 maps to tb_1 and X_2 to $-tb_2$, so that $\mathfrak{P}'$ is prime and has zero intersection with Q. Hence $(b_1X_1+b_2X_2)Q_g$ is a prime ideal meeting Q in the zero ideal, so that X(I) = (0), but there are prime ideals of the form $\mathfrak{P}_i\cap Q$ which are not zero, namely the prime ideals minimal over aQ.

3. Some invariants of sets of ideals of a local ring.

In this section, we review the results and definitions of chapter 10 and the first two sections of the present chapter and bring together a number of these results and definitions to make reference back in the next chapter somewhat easier.

First, we consider a set of ideals $\mathbf{I} = (I_1,\ldots,I_s)$ and an independent set of general elements $(x_1,\ldots,x_s)$ of this set. Then it has already been observed, in Corollaries ii) and iii) of Theorem 11.11, that the ring

$$L(I) = Q_g/(x_1,\ldots,x_s)Q_g$$

considered as a Q-algebra, and the ideal

$$X(I) = (x_1,\ldots,x_s)Q_g\cap Q$$

of Q, are both independent of the choice of the set of independent general elements $(x_1,\ldots,x_s)$. We can therefore consider them as invariants of the set of ideals I. From these basic invariants we can construct others. For example, consider the set of prime ideals $\mathfrak{P}_1,\ldots,\mathfrak{P}_n$ of Q_g minimal over the ideal $(x_1,\ldots,x_s)Q_g$. This set is not an invariant in the sense just described, but the set of prime ideals $\mathfrak{p}_i = \mathfrak{P}_i \cap Q$, $i = 1,\ldots,n$ is an invariant. It is worthwhile giving these ideals a name.

DEFINITION. *The prime ideals* $\mathfrak{p}_1,\ldots,\mathfrak{p}_n$ *will be termed the limit ideals of* I.

Now we refer back to Theorem 11.21. The prime ideals $\mathfrak{P}_i$ are clearly in 1-1 correspondence with the minimal prime ideals of L(I), while the correspondence between the sets $\mathfrak{P}_1,\ldots,\mathfrak{P}_n$ and $\mathfrak{p}_1,\ldots,\mathfrak{p}_n$ is 1-1 in view of 11.21 iii). Hence we have a 1-1 correspondence between the limit ideals $\mathfrak{p}_i$ and the minimal prime ideals of the ring L(I). We will denote the minimal prime ideal of L(I) corresponding to $\mathfrak{p}_i$ by $\mathfrak{P}'_i$.

DEFINITION. *If* $\mathfrak{p}_i$ *is a limit ideal of* I, *the quotient algebra* $L(I)/\mathfrak{P}'_i$, *where* $\mathfrak{P}'_i$ *is as just defined, will be denoted by* $L(I,\mathfrak{p}_i)$.

Note that $L(I,\mathfrak{p}_i)$ is a quotient algebra of Q_g, and, as such, is the general extension of the ring $Q_N/\mathfrak{P}_i \cap Q_N$, where $N \geq \mathrm{def}(\mathfrak{P}_i)$.

Now we consider the implications of 11.21 ii). This associated with each of the ideals $\mathfrak{P}_i$ an integer $t(\mathfrak{P}_i)$. In view of the 1-1 correspondence $\mathfrak{p}_i \leftrightarrow \mathfrak{P}_i$ we associate this integer with $\mathfrak{p}_i$ and write it as $t(\mathfrak{p}_i)$. We can use 11.21 ii) to define $t(\mathfrak{p}_i)$ directly in terms of $\mathfrak{p}_i$.

DEFINITION. *If* $\mathfrak{p}$ *is a limit ideal of the set of ideals* I, *then* $t(\mathfrak{p})$ *is defined to be the number of ideals* I_j *not contained in* $\mathfrak{p}$. *The set of ideals* I_j *contained in* $\mathfrak{p}$ *will be denoted by* $I'(\mathfrak{p})$ *and the set of ideals* $I_j + \mathfrak{p}/\mathfrak{p}$, *where* I_j *ranges over the set* $I - I'(\mathfrak{p})$, *will be denoted by* $I''(\mathfrak{p})$.

We now consider the three Theorems 10.23, 11.21, and 11.25, which will play an important role in the next chapter. Their statements are closely related, while the hypotheses on the ideals concerned in these theorems differ slightly in each case. We therefore conclude this chapter with an omnibus theorem which includes the results of the three theorems cited and the conditions under which the various statements are proved. It is convenient first of all to introduce a further definition.

DEFINITION. *A limit ideal* $\mathfrak{p}$ *of a set of ideals* $I = (I_1,\dots,I_s)$ *will be termed acceptable, if* $\mathfrak{p}$ *is the intersection with* Q *of a good minimal prime ideal* $\mathfrak{P}$ *of* $(x_1,\dots,x_s)Q_g$ *for some independent set of general elements* $(x_1,\dots,x_s)$ *of* I.

THEOREM 11.31. *Let* $(Q,\mathfrak{m},k,d)$ *be a local ring,* $\mathfrak{p}$ *be a prime ideal of* Q. *Let* Q^* *denote either* Q_g *or* Q_N, *and let* $\mathfrak{P}$ *be a prime ideal of* Q^* *meeting* Q *in* $\mathfrak{p}$. *Let* F *denote the field of fractions of* $Q/\mathfrak{p}$ *and* E *the field of fractions of* $Q^*/\mathfrak{P}$.

i) a) $R^* = (Q^*)_{\mathfrak{P}}/\mathfrak{p}(Q^*)_{\mathfrak{P}}$ *is regular of dimension* $t = t(\mathfrak{P})$, *where* $t(\mathfrak{P})$ *can be defined either as* $\operatorname{ht}\mathfrak{P} - \operatorname{ht}\mathfrak{p}$ *or, in the case where* $Q^* = Q_N$, *by the equation*

$$\text{trans.deg}_F E = N - t(\mathfrak{P}),$$

b) $\mathfrak{p}Q^*_{\mathfrak{P}}$ *is a good prime ideal of* $Q^*_{\mathfrak{P}}$,

c) $\dim(Q/\mathfrak{p}) - \dim(Q^*/\mathfrak{P}) \geq t(\mathfrak{P})$.

ii) *Let* $I = (I_1,\dots,I_s)$ *be a set of ideals of* Q, *let* $N \geq \operatorname{def}(I)$, *let* $\mathfrak{p}$ *be a limit ideal of* I, *and let* $(x_1,\dots,x_s)$ *be an independent set of general elements of* I *in* Q^*. *Then*

a) *there is exactly one prime ideal* $\mathfrak{P}$ *of* Q^* *minimal over* $(x_1,\dots,x_s)Q^*$ *meeting* Q *in* $\mathfrak{p}$, *and* $\mathfrak{p}Q^* = \mathfrak{P} \cap \mathfrak{a}$, *where* $\mathfrak{a}$ *contains elements of Q not in* $\mathfrak{p}$,

b) $\mathfrak{p}$ *contains exactly* $s - t(\mathfrak{P})$ *of the ideals* I_j *and, if* I *is so numbered that these are* $I_1,\dots,I_{s-t}$, *with the corresponding renumbering of the elements* $x_1,\dots,x_s$, *then the images of* $x_{s-t+1},\dots,x_s$ *in* R^* *form a minimal basis of the maximal ideal* $\mathfrak{n}$ *of* R^*. *Further,* $t(\mathfrak{P})$ *depends only on* $\mathfrak{p}$ *and can be written as* $t(\mathfrak{p})$.

iii) *Suppose that, in* ii), $\mathfrak{p}$ *is assumed to be acceptable, i.e.,* $\mathfrak{P}$ *is good. Then*

a) $\mathfrak{p}$ *has height* $s - t(\mathfrak{p})$ *and* $Q/\mathfrak{p}$ *has dimension* $d - s + t(\mathfrak{p})$,

b) $Q^*/\mathfrak{P}$ *has dimension* $d - s$,

c) *if* $I'(\mathfrak{p})$ *is the sub-set of* I *consisting of those* I_j *contained in* $\mathfrak{p}$ *and* $\mathbf{X}'$ *denotes any independent set of general elements of the set of ideals* $I'(\mathfrak{p})$

contained in Q^*, *then* $\mathbf{p}Q^*$ *is minimal over* $\mathbf{x}'Q^*$.

d) $\mathbf{p}Q^*$ *is the only minimal prime ideal of* $\mathbf{x}'Q^*$ *such that* $(x_1,\ldots,x_{s-t})Q^*_{\mathbf{p}}$ *has radical* $\mathbf{p}Q^*_{\mathbf{p}}$.

i) a) is given in Theorem 10.23, the first of the alternative definitions of $t(\mathbf{p})$ being given by the equality in Theorem 10.23 i), and the second being the original definition. i) b) is a restatement of the equality of the first definition, given the fact that $\dim R^* = t(\mathbf{p})$. i) c) is stated in 10.23 i).

ii) a) and ii) b) are restatements of 11.21.

Finally iii) a)-d) are restatements of 11.25.

12. MIXED MULTIPLICITIES AND THE GENERALISED DEGREE FORMULA

1. Multiplicities again.

In this chapter we will return to the notion of the multiplicity function associated with an $\mathfrak{m}$-primary ideal $(a_1,\dots,a_s)$ of a local ring $(Q,\mathfrak{m},k,d)$. The most convenient approach for our purposes in the present chapter is the inductive approach due to D.J. Wright, and described in detail in Northcott's book [LRMM]. We therefore begin this section with a description of this approach, referring to [LRMM] for the proofs. In order to avoid difficulties arising from the use of multiple suffixes, we make a slight change in notation below. Where Wright and Northcott have used $e_Q(a_1,\dots,a_s \mid M)$ we will use $e(Q\mid a_1,\dots,a_s \mid M)$. We will also write $e(Q\mid a_1,\dots,a_s)$ for $e(Q\mid a_1,\dots,a_s \mid Q)$.

We take $(Q,\mathfrak{m},k,d)$ *and* $(a_1,\dots,a_s)$ *as above and suppose that* M *is any finitely generated* Q*-module. We now define* $e(Q\mid a_1,\dots,a_s \mid M)$ *by induction on* s. *First, if* $s = 0$, *so that* Q *is artinian, we define* $e(Q\mid . \mid M)$ *to be the length* $l_Q(M)$ *of the* Q*-module* M. *If* $s > 0$, *write* Q' *for the local ring* Q/a_1Q *and consider* M/a_1M *and* $(0 : a_1)M$ *as* Q'*-modules. Let* a'_1 *be the image of* a_1 *in* Q'. *Then we define*

$$e(Q\mid a_1,\dots,a_s \mid M) = e(Q'\mid a'_2,\dots,a'_s \mid M/a_1M) - e(Q'\mid a'_2,\dots,a'_s \mid (0 : a_1)_M).$$

This is, in essence, the definition of [LRMM] p.299.

We now recall a number of basic results from [LRMM]. We will abbreviate $(a_1,\dots a_s)$ to $\mathbf{a}$.

(12.1.1). *If*

$$0 \longrightarrow M' \longrightarrow M \longrightarrow M'' \longrightarrow 0$$

is an exact sequence of Q*-modules, then*

$$e(Q\mid \mathbf{a} \mid M) = e(Q\mid \mathbf{a} \mid M') + e(Q\mid \mathbf{a} \mid M'').$$

[LRMM] p.302, Theorem 5.

(12.1.2). *If* π *is a permutation of* 1 *to* s, *and* $\pi(\mathbf{a})$ *denotes the sequence* $a_{\pi(1)},\dots,a_{\pi(s)}$, *then*

$$e(Q|\ \pi(\mathbf{a})\ |M) = e(Q|\ \underline{a}\ |M).$$

[LRMM] p.306, Proposition 4.

(12.1.3). $0 \le e(Q|\ \mathbf{a}\ |M) \le l_Q(M/a_1M + \dots + a_sM)$.

[LRMM] p.308, Theorem 6.

(12.1.4). *If* $a_k = a'_k a''_k$, *then* $a_1Q + \dots + a_sQ$ *is* $\mathfrak{m}$*-primary if and only if both* $a_1Q + \dots + a'_kQ + \dots + a_sQ$ *and* $a_1Q + \dots + a''_kQ + \dots + a_sQ$ *are* $\mathfrak{m}$*-primary, and, when this is the case,*

$$e(Q|\ a_1,\dots,a_s\ |M) = e(Q|\ a_1,\dots,a'_k,\dots,a_s\ |M) + e(Q|\ a_1,\dots,a''_k,\dots,a_s\ |M).$$

[LRMM] p.309, Theorem 7.

(12.1.5). *If* $(b_1,\dots,b_s)$ *is a second set of elements of* Q *such that* $b_1Q+\dots+b_sQ \supseteq a_1Q+\dots+a_sQ$, *then, for all* M,

$$e(Q|\ b_1,\dots,b_s\ |M) \le e(Q|\ a_1,\dots,a_s\ |M).$$

[LRMM] p.330, Theorem 14.

(12.1.6). *If* $s > d$, *then* $e(Q|\ \mathbf{a}\ |\ M) = 0$, *for all* M, *while if* $s = d$, $e(Q|\ \mathbf{a}\ |Q) > 0$.

[LRMM] p.334 Proposition 7.

Before our next result we make some observations. The set of elements $(a_1,\dots,a_s)$ has been considered as an ordered set, the ordering being inherent in the use of the set 1,2,...,s as a set of suffixes, and the ordering was used in the inductive definition of $e(Q|\ \mathbf{a}\ |M)$. However 12.1.2 implies that the value of $e(Q|\ \mathbf{a}\ |M)$ is independent of the ordering. Hence in the theorem following, we will consider $\mathbf{a}$ as a set indexed by an unordered set S containing s elements, and we will write it as $\mathbf{a}_S$ to indicate that this is the case. If U is a sub-set of S, then we will write $\mathbf{a}_U$ for the set of elements a_i with $i \in U$. We will also write $\mathbf{a}_UQ$ for the ideal generated by the elements of $\mathbf{a}_U$. As usual, S-U will denote the set of elements of S not in U. The number of elements in U will be denoted by $|U|$. Finally, if $\boldsymbol{\theta}$ is a function on a set S and $\mathbf{x}$ is a set of elements of S, $\theta(\mathbf{x})$ will denote the set of elements $\theta(x)$, where x ranges over $\mathbf{x}$.

(12.1.7). *Let* U *be a sub-set of* S *and let* $\mathbf{p}$ *range over all prime ideals minimal over* $\mathbf{a}_U Q$. *Let* $\phi_{\mathbf{p}}$ *denote the canonical map of* Q *into* $Q/\mathbf{p}$ *and* $\theta_{\mathbf{p}}$ *denote the canonical map of* Q *into* $Q_{\mathbf{p}}$. *Then*

$$e(Q|\,\mathbf{a}_S\,|M) = \Sigma e(Q_{\mathbf{p}}|\,\theta_{\mathbf{p}}(\mathbf{a}_U)\,|M_{\mathbf{p}})e(Q/\mathbf{p}\,|\,\phi_{\mathbf{p}}(\mathbf{a}_{S-U})\,|Q/\mathbf{p})$$

where the sum is over all prime ideals $\mathbf{p}$ *minimal over* $\mathbf{a}_U Q$.

(12.1.7) is the associativity law for multiplicities ([LRMM], p.342, Theorem 18). We will also use a slightly modified form. If the prime ideal $\mathbf{p}$ is not good, then either $\dim Q_{\mathbf{p}} < s$ or $\dim(Q/\mathbf{p}) < s - t$. In each case 12.1.6 then implies that the term in the above sum corresponding to $\mathbf{p}$ is zero. Hence we can restrict the above sum to those $\mathbf{p}$ which are good prime ideals minimal over $\mathbf{a}_U Q$.

We will now give a proof of the result mentioned before Theorem 11.25 in the last chapter. The result is a Corollary of 12.1.7 and is given as such.

COROLLARY. *Let* $(Q,\mathfrak{m},k,d)$ *be a local ring and let* $\mathbf{a}_S = (a_1,\dots,a_d)$ *be a set of parameters of* Q. *Then if* U *is a sub-set of the set* $S = (1,\dots,d)$, *there is a good prime ideal* $\mathbf{p}$ *minimal over* $\mathbf{a}_U Q$.

We apply the modfied form of 12.1.7 to S,U, taking M = Q. By 12.1.6, the left-hand side is non-zero, implying that there is at least one non-zero term on the right-hand side, and hence at least one good prime ideal minimal over $\mathbf{a}_U Q$.

THEOREM 12.1.8. *Let* $(Q,\mathfrak{m},k)$ *be a 1-dimensional local domain with field of fractions* F. *Then there exists a finite set of valuations* $v_1,\dots,v_s$ *on* F, *satisfying* $v_i(x) \geq 0$ *on* Q *and* >0 *on* $\mathfrak{m}$, *and integers* δ_i *such that, for any non-zero* a *of* Q,

$$e(Q|\,a\,|Q) = \Sigma\delta_i[k_i;k]v_i(a)$$

where k_i *is the residue field of* v_i.

We commence with the observation that

$$e(Q| a | Q) = l_Q(Q/aQ)$$

since, as Q is a domain, a is not a zero divisor.

Now suppose we replace Q by its completion Q^. Then a is not a zero divisor in Q^, and hence $e(Q^\wedge| a |Q^\wedge) = l_{Q^\wedge}(Q^\wedge/aQ^\wedge) = l_Q(Q/aQ) = e(Q| a |Q)$.

Now let $\mathbf{p}_1,\dots,\mathbf{p}_s$ be the minimal prime ideals of Q^, and let δ_i be the length of the corresponding isolated component of zero in Q^. Then

$$e(Q^\wedge| a | Q^\wedge) = \Sigma\delta_i e(Q^\wedge| a |Q^\wedge/\mathbf{p}_i) \qquad (12.1.8.1)$$

summed from i = 1 to s. Now let Q' denote one of the rings $Q^\wedge/\mathbf{p}_i$ so that Q' is a complete local domain of dimension 1. Then we can write $e(Q^\wedge| a |Q')$ as $e(Q'| a' |Q')$. Now suppose b' is any non-zero element of Q'. Then, if $f = f_{b'Q'}$,

$$d(f,a') = l(Q'/a'Q') = e(Q'| a' |Q'),$$

by the definition of degree function at the beginning of chapter 9, noting that

$$(0 : a)_{Q'} = (0).$$

By Lemma 9.36, it follows that

$$e(Q'| a' |Q') = d(f,a') = \Sigma[k_j : k]v'_j(a')$$

the sum being over the set of valuations v'_j on the field of fractions of Q' which are ≥0 on Q' and >0 on $\mathbf{m}'$. Substituting in (12.1.8.1), and replacing the valuations v'_j by their restrictions to the field of fractions of Q and $e(Q^\wedge| a | Q^\wedge)$ by $e(Q| a |Q)$, we obtain an equation of the given form.

We note that Q can be considered as a sub-ring of both Q_N and Q_g, and Q_N can be considered as a sub-ring of Q_g. Hence we will consider elements of Q as elements of either Q_N or Q_g and elements of Q_N as elements of Q_g without further comment.

LEMMA 12.1.9. *Let* $(Q,\mathbf{m},k,d)$ *be a local ring,* M *be a finitely generated* Q*-module and let* $\mathbf{a} = (a_1,\dots,a_s)$ *be a set of elements of* Q *generating an* $\mathbf{m}$*-primary ideal. Let* Q^* *denote either* Q_g *or* Q_N *and let* M^* *denote* M_g *if* $Q^* = Q_g$ *and* M_N *if* $Q^* = Q_N$. *Then,*

$$e(Q^*| \mathbf{a} |M^*) = e(Q| \mathbf{a} |M)$$

First we can impose the restriction that s = d, since if s > d, both sides of the equation are zero as $\dim Q^* = \dim Q = d$.

We consider the two cases together and proceed by induction on d. If d = 0, the result follows since, for any finitely generated Q-module M, $l_{Q*}(M*) = l_Q(M)$ by (10.11) and (7.41). If d = 1, then

$$l(M*/a_1M*) = l((M/a_1M)*) = l(M/a_1M)$$

and

$$l((0 : a_1)_{M*}) = l(((0 : a_1)_M)*) = l((0 : a_1)_M)$$

proving the equality in the case d = 1. The inductive step follows in almost exactly the same way as the case d = 1.

Before the second of our lemmas we need to add to our list of elementary results.

(12.1.10) *Let* (R,$\mathfrak{m}$,k,d) *be a regular local ring of dimension* d *and let* $\underline{a} = (a_1,\ldots,a_d)$ *be a minimal set of generators of* $\mathfrak{m}$. *Then*

$$e(R \mid \mathbf{a}) = 1.$$

The proof is by induction on d, the case d = 0 being trivial. We use the fact that $R/a_1R + \ldots + a_iR$ is regular for i = 1,...,d.

LEMMA 12.1.11. *Let* (Q,$\mathfrak{m}$,k,d) *be a local ring,* $\mathbf{a} = (a_1,\ldots,a_s)$ *be a set of elements of* $\mathfrak{m}$, *generating an* $\mathfrak{m}$*-primary ideal and suppose that, for some* u, $1 \leq u \leq s$, *the radical of the ideal* $(a_1,\ldots,a_u)$ *is a prime ideal* $\mathfrak{p}$ *such that* $Q/\mathfrak{p}$ *is regular of dimension* u, *and that the set of images* $\phi_{\mathfrak{p}}(a_j)$ *of* a_j *in* $Q/\mathfrak{p}$, j = u + 1,...,s, *form a minimal basis of* $\mathfrak{m}/\mathfrak{p}$. *Then, if* M *is any finitely generated* Q*-module,*

$$e(Q \mid \mathbf{a} \mid M) = e(Q_{\mathfrak{p}} \mid \theta_{\mathfrak{p}}(a_1),\ldots, \theta_{\mathfrak{p}}(a_u) \mid M_{\mathfrak{p}}).$$

This follows immediately from (12.1.7) with U = (1,...,u), noting that the second factor in the single term is 1 by (12.1.10).

2. Mixed multiplicities.

DEFINITION. *Let* (Q,$\mathfrak{m}$,k,d) *be a local ring and* $\mathbf{I} = (I_1,\ldots,I_s)$ *be a set of ideals of* Q *such that if* $x_1,\ldots,x_s$ *is an independent set of general elements of* $I_1,\ldots,I_s$, *the ideal* $(x_1,\ldots,x_s)$ *is* $\mathfrak{m}$*-primary. Then, if* M *ranges over the set of finitely generated* Q*-modules,* *we define the mixed multiplicity function* $e(Q|I|M)$ *to be* $e(Q_g \mid x_1,\ldots,x_s \mid M_g)$.

DEFINITION. *A set of ideals* I *as above with the property that the ideals generated by some, and hence all, independent sets of general elements of* I *are* $\mathfrak{m}$*-primary will be termed a large set of ideals. (Note that this implies that* I *contains at least* d *members and if it contains exactly* d *members, it is large if and only if it is an independent set of ideals.)*

The first definition, due essentially to Teissier[1973], is independent of the choice of the independent set of general elements $x_1,\ldots,x_s$, by Corollary ii) to Theorem 11.11. This is the main reason for phrasing the definition in terms of Q_g and M_g. However it will often be convenient below to use an alternative definition in which Q_g, M_g are replaced by Q_N, M_N where N is large enough. The phrase "N is large enough" here means "Q_N contains the general elements $x_1,\ldots,x_s$". Note that it is an immediate consequence of Lemma 12.1.9 that if N is large enough in this sense, then

$$e(Q|I|M) = e(Q_g|x_1,\ldots,x_s|M_g) = e(Q_N|x_1,\ldots,x_s|M_N),$$

since Q_g is isomorphic to $(Q_N)_g$.

The case $s = d = 1$ is important in inductions and in this case it is possible to define $e(Q|I|M)$ without introducing general extensions as is shown by the following theorem, designated Theorem 12.2A in order not to interfere with the correspondence between the numbering of the results that follow and results of section 12.1.

THEOREM 12.2A. *Let* $(Q,\mathfrak{m},k)$ *be a 1-dimensional local ring, let* I *be an* $\mathfrak{m}$*-primary ideal of* Q*, and let* M *be any finitely generated* Q*-module. Then,*

i) *for large* n*, and* K' *a constant,*

$$l_Q(M/I^n) = e(Q|I|M)n + K'$$

ii) *if* x *is any element of* Q_g *such that* xQ_g *is a reduction of* IQ_g*, then*

$$e(Q|I|M) = e(Q_g|x|M).$$

i) Let x be an element of I such that xQ is a reduction of I. Then, by the definition of $e(Q|\mathbf{x}|M)$ and 12.1.4

$$ne(Q|x|M) = e(Q|x^n|M) = l_Q(M/x^nM) - l_Q((0:x^n)_M).$$

Now the last term increases with n to a constant value K, independent of x. Hence

$$l_Q(M/x^nM) = ne(Q\,|\,x\,|\,M) + K \quad \text{for large n.}$$

As x is a reduction of the ideal I, there exists q such that

$$xI^q = I^{q+1},$$

whence, for all $n \geq q$, $x^{n-q}I^q = I^n$. This implies that, if $n \geq q$,

$$l_Q(I^nM/x^nM) \geq l(x.I^nM/x.x^nM) \geq l_Q(I^{n+1}M/x^{n+1}M),$$

from which it follows, that, for n large, $l_Q(I^nM/x^nM)$ is constant and hence

$$l_Q(M/I^nM) = e(Q\,|\,x\,|\,M)n + K', \quad \text{where } K' \text{ is a constant.}$$

Now as Q is of dimension 1, I has analytic spread at most 1 and hence equal to 1. Hence it follows that, if x is a general element of I, xQ_g is a reduction of IQ_g, by 11.22. Hence for large n, the length of M_g/I^nM_g, considered as a Q_g-module, is equal to $e(Q_g\,|\,x\,|\,M_g)n + K''$, where K" is a constant. But $e(Q_g\,|\,x\,|\,M_g)$ is, by definition, equal to $e(Q\,|\,I\,|\,M)$, while, as general extension preserves lengths, $l(M_g/I^nM_g) = l_Q(M/I^nM)$, and the result follows, with $K' = K''$.

ii) The last two sentences above show that, with such an x, the length of the Q_g-module M_g/I^nM_g is equal to $e(Q_g\,|\,x\,|\,M_g)n$ + constant for large n. Since i) shows that it is also equal to $e(Q\,|\,I\,|\,M)n$ + constant for large n, we have

$$e(Q_g\,|\,x\,|\,M_g) = e(Q\,|\,I\,|\,M).$$

We now translate the results of the last section to yield results on $e(Q\,|\,I\,|\,M)$. We give these results without comment, except where the proofs are not immediate.

(12.2.1) *If*

$$0 \longrightarrow M' \longrightarrow M \longrightarrow M'' \longrightarrow 0$$

is an exact sequence of Q-modules, then

$$e(Q\,|\,I\,|\,M) = e(Q\,|\,I\,|\,M') + e(Q\,|\,I\,|\,M'').$$

(12.2.2) *If* π *is a permutation of* 1 *to* s, *and* $\pi(I)$ *denotes the sequence of ideals* $I_{\pi(1)},\ldots,I_{\pi(s)}$, *then*

$$e(Q\,|\,\pi(I)\,|\,M) = e(Q\,|\,I\,|\,M).$$

(12.2.3) $0 \leq e(Q\,|\,I\,|\,M) \leq l(M_g/x_1M_g + \ldots + x_sM_g)$, *where* $x_1,\ldots,x_s$ *is an independent set of general elements of* I, *the length in the right-hand term being as a* Q_g*-module.*

Note that this length depends only on M, I, and not on the choice of the general elements $x_1,\ldots,x_s$.

The proof of our next two statements will require reduction to the case $d = 1$, and they will be proved after the statement of (12.2.6).

(12.2.4) *If* $I = (I_1,\ldots,I_s)$ *and* $I_k = I'_k I''_k$, *then* I *is large if and only if both the sets* I', I'' *obtained from* I *by replacing* I_k *by* I'_k *or* I''_k *respectively are large and, when this is the case,*

$$e(Q\,|\,I\,|\,M) = e(Q\,|\,I'\,|\,M) + e(Q\,|\,I''\,|\,M).$$

(12.2.5) *If* I', I'' *are two large sets of* s *ideals such that* $I'_j \supseteq I''_j$ *for* $j = 1,\ldots,s$, *then*

$$e(Q\,|\,I'\,|\,M) \leq e(Q\,|\,I''\,|\,M)$$

for all M.

(12.2.6) *If* $s > d$, *then* $e(Q\,|\,I\,|\,M) = 0$, *for all* M, *while if* $s = d$, $e(Q\,|\,I\,|\,M) > 0$.

This follows immediately from (12.1.6).

We now consider the proof of the statements (12.2.4) and (12.2.5). In view of (12.2.6), the last statements of (12.2.4) and (12.2.5) are trivial if $s > d$, and we therefore restrict attention to the case $s = d$.

First suppose that $d = 1$. Let I be an $\mathfrak{m}$-primary ideal of Q. We use the alternative definition of $e(Q\,|\,I\,|\,M)$ given by Theorem 11.2A.

First we note that if I' is a second $\mathfrak{m}$-primary ideal of Q such that $I \supseteq I'$, then Theorem 12.2A i) implies that

a) $e(Q\,|\,I\,|\,M) \leq e(Q\,|\,I'\,|\,M)$.

Secondly, we note that, if I, I' are two $\mathfrak{m}$-primary ideals and xQ, x'Q are reductions of I, I' respectively, then xx'Q is a reduction of II'. Then Theorem 12.2A ii) and (12.1.4) imply that

b) $e(Q\,|\,II'\,|\,M) = e(Q\,|\,I\,|\,M) + e(Q\,|\,I'\,|\,M)$.

We can now give the basic lemma for the proof of (12.2.4) and (12.2.5).

LEMMA. *If* $e(Q\,|\,I_1,\dots,I_d\,|\,M)$ *is the mixed multiplicity function defined above, then*

i) $e(Q\,|\,I_1,\dots,I_{d-1},I'_dI''_d\,|\,M) = e(Q\,|\,I_1,\dots,I_{d-1},I'_d\,|\,M) + e(Q\,|\,I_1,\dots,I_{d-1},I''_d\,|\,M)$,

ii) *if* $I'_d \supseteq I''_d$, *then* $e(Q\,|\,I_1,\dots,I_{d-1},I'_d\,|\,M) \le e(Q\,|\,I_1,\dots,I_{d-1},I''_d\,|\,M)$.

The case d = 1 of both i) and ii) follow from b), a) above. The case d > 1 follows from the associativity formula (12.1.7) applied with U = (1,2,...,d−1), which gives

$$e(Q\,|\,I_1,\dots,I_d\,|\,M) = \Sigma e((Q_g)_{\mathfrak{p}}\,|\,\theta_{\mathfrak{p}}(I_1),\dots,\theta_{\mathfrak{p}}(I_{d-1})\,|\,(M_g)_{\mathfrak{p}})e(Q_g/\mathfrak{p}\,|\,\phi_{\mathfrak{p}}(I_d)\,|\,Q_g/\mathfrak{p})$$

the sum being over all good minimal prime ideals $\mathfrak{p}$ over $x_1Q_g + \dots + x_{d-1}Q_g$. We simply apply the case d = 1 to the second factor of each term in the sum, Q being taken as $Q_g/(x_1Q_g + \dots + x_{d-1}Q_g)$.

Now we note that (12.2.4) follows from part i) and (12.2.1), while (12.2.5) follows by (12.2.1) and repeated application of part ii).

For the moment, we skip the generalisation of (12.1.7) and proceed to the generalisations of (12.1.8) and (12.1.9).

(12.2.8) *Let* $(Q,\mathfrak{m},k)$ *be a 1-dimensional local domain with field of fractions* F. *Then there exists a finite set of valuations* $v_1,\dots,v_s$ *on* F, *satisfying* $v_i(x) \ge 0$ *on* Q *and* >0 *on* $\mathfrak{m}$, *and integers* δ_i *such that, for any* $\mathfrak{m}$*-primary ideal* I *of* Q,

$$e(Q\,|\,I\,|\,Q) = \Sigma\delta_i[k_i : k]v_i(I),$$

where k_i *is the residue field of* Q.

By definition, $e(Q\,|\,I\,|\,Q) = e(Q_g\,|\,x\,|\,Q_g)$, where x is a general element of I. Hence, by (12.1.8) there exists a finite set of valuations $V_1,\dots,V_s$ on F_g, ≥0 on Q_g, >0 on m_g and integers $\delta_1,\dots,\delta_s$ such that

$$e(Q\,|\,I\,|Q) = \Sigma\delta_i[K_i:k_g]V_i(x),$$

where K_i is the residue field of V_i. Now K_i is algebraic over k_g. Hence it follows from Lemma 10.33 that, if v_i is the restriction of V_i to F, then V_i is the general extension of v_i. This further implies that, if k_i is the residue field of v_i, then $K_i = k_{ig}$ and hence $[K_i : k_g] = [k_i : k]$. Finally, xQ_g is a reduction of IQ_g by 11.22, which implies that

$V_1(x) = V_1(IQ_g) = v_1(I)$. Hence, making the appropriate substitutions, we obtain (12.2.8).

(12.2.9) *Let* $(Q,\mathfrak{m},k,d)$ *be a local ring,* M *be a finitely generated* Q-*module and let* $I = (I_1,\ldots,I_s)$ *be a large set of ideals of* Q. *Let* Q* *denote either* Q_g or Q_N *and let* M* *denote* M_g *if* $Q^* = Q_g$ *and* M_N *if* $Q^* = Q_N$. *Then,*

$$e(Q^* \mid IQ^* \mid M^*) = e(Q \mid I \mid M).$$

We can restrict attention to the case s = d. Let $\mathbf{x} = (x_1,\ldots,x_d)$ be any set of general elements of **I**, and let N' ≥ def**x**. Then we have already remarked that

$$e(Q \mid I \mid M) = e(Q_{N'} \mid \mathbf{x} \mid M_{N'}).$$

First take $Q^* = Q_N$. Then, by applying to Q_g an automorphism which increases the suffixes of the X_i involved in the expressions for $x_1,\ldots,x_d$ by N and replacing N' by N+N' we may assume that **x** is also a general set of elements of IQ_N. Then

$$e(Q_N \mid IQ_N \mid M_N) = e(Q_{N+N'} \mid \mathbf{x} \mid M_{N+N'}) = e(Q \mid I \mid M).$$

Now suppose that $Q^* = Q_g$. By definition, $e(Q_g \mid IQ_g \mid M_g) = e((Q_g)_g \mid \mathbf{x} \mid (M_g)_g)$, where **x** is any general set of elements of IQ_g contained in $(Q_g)_g$. We will take $(Q_g)_g$ to be the localisation of $Q_g[Y_1,Y_2,\ldots]$ at $\mathfrak{m}_g[Y_1,Y_2,\ldots]$. By taking **x** to be standard, we can assume that the elements $x_1,\ldots,x_d$ are contained in $Q[Y_1,Y_2,\ldots,Y_N]$ for some N. Then

$$e(Q_g \mid IQ_g \mid M_g) = e((Q_g)_N \mid \mathbf{x} \mid (M_g)_N).$$

But we can replace $(Q_g)_N$ and $(M_g)_N$ by $(Q_N)_g$ and $(M_N)_g$ on the right-hand side. Applying (12.1.8) we can reduce the right-hand side to $e(Q_N \mid \mathbf{x} \mid M_N)$ which finally, by the remark already made, is $e(Q \mid I \mid M)$.

Now we turn attention to the more difficult task of generalising (12.1.7).

We restrict attention to the case where s = d, and take $I_S = (I_1,\ldots,I_d)$ to be a large (or, equivalently, an independent) set of d ideals of Q so that if $\mathbf{x}_S = (x_1,\ldots,x_d)$ is an independent set of general elements of I_S, then the ideal $(x_1,\ldots,x_d)Q_g$ is $\mathfrak{m}_g$-primary. Let U be a sub-set of S, where u = |U| and 1 ≤ u ≤ d. By suitable re-numbering we can take U to be the set (1,...,u). Write $\mathbf{x}_U$ for the set (1,...,u). Then a direct application of (12.1.7) gives

$$e(Q \mid I_S \mid M) = \Sigma_{\mathcal{P}} e((Q_g)_{\mathcal{P}} \mid \theta_{\mathcal{P}}(x_U) \mid (M_g)_{\mathcal{P}}) e(Q_g/\mathcal{P} \mid \phi_{\mathcal{P}}(x_{S-U}) \mid Q_g/\mathcal{P}) \qquad (*),$$

the sum being over all good primes $\mathcal{P}$ of Q_g minimal over $x_U Q_g$. However, the formula above essentially involves the choice of the general elements x_S. In particular, this choice determines the set of prime ideals $\mathcal{P}$ over which we sum. Our aim is to modify the terms on the right-hand side of (*) so as to remove any reference to a particular set of independent general elements of I.

We commence with the observation, already made in section 11.3 that, whereas the set of ideals $\mathcal{P}$ does depend on the choice of x_S, the set of ideals $\mathbf{p} = \mathcal{P} \cap Q$ is the same for all choices of x_S (or, strictly speaking x_U), and, further, for a particular choice, the correspondence $\mathbf{p} \leftrightarrow \mathcal{P}$ is 1-1. Hence we shall consider the sum on the right-hand side as being over all acceptable limit ideals $\mathbf{p}$ of **I**, these being the ideals $\mathbf{p}$ corresponding to the minimal prime ideals of $x_U Q_g$ which are good. We now aim at expressing the term on the right-hand side in terms of the corresponding $\mathbf{p}$ rather than, as above, $\mathcal{P}$. We now make a second observation. The ring $Q_g/\mathcal{P}$ is, to within isomorphism as a Q-algebra, determined by I_U and $\mathbf{p}$ and will be denoted by $L(I_U, \mathbf{p})$ as in section 11.3.

With this in mind, we now look at the first factor of a term on the right-hand side of (*). Now consider the subset I'_U of I_U consisting of those ideals I_j contained in $\mathbf{p}$. We will, renumbering if necessary, suppose these are $I_1,\dots,I_{u-t}$, where $t = t(\mathbf{p})$. We now recall that,

1) by 11.31 i) a), $R(\mathcal{P}) = (Q_g)_{\mathcal{P}}/\mathbf{p}(Q_g)_{\mathcal{P}}$ is regular of dimension t, and that its maximal ideal is generated by the images of $x_{d-t+1},\dots,x_d$, and,

2) by 11.31 iii), $\mathbf{p}(Q_g)_{\mathcal{P}}$ is the only minimal prime ideal of the ideal $x_1(Q_g)_{\mathcal{P}} + \dots + x_{d-t}(Q_g)_{\mathcal{P}}$.

Hence we can apply 12.1.7 to the term

$$e((Q_g)_{\mathcal{P}} \mid \theta_{\mathcal{P}}(x_U) \mid (M_g)_{\mathcal{P}})$$

and writing $\mathcal{P}^*$ for $\mathbf{p}(Q_g)_{\mathcal{P}}$ for typographical reasons, we obtain as equal to it,

$$e(((Q_g)_{\mathfrak{p}})_{\mathfrak{p}^*} \mid \theta_{\mathfrak{p}}((x_1,\ldots,x_{d-t})) \mid ((M_g)_{\mathfrak{p}})_{\mathfrak{p}^*})\ e(R(\mathfrak{P}) \mid \phi(x_{d-t+1}),\ldots,\phi(x_d) \mid R(\mathfrak{P})).$$

Now 12.1.10 implies that the second factor is simply 1, while, since $((Q_g)_{\mathfrak{p}})_{\mathfrak{p}^*}$ is the same as $(Q_{\mathfrak{p}})_g$, $((M_g)_{\mathfrak{p}})_{\mathfrak{p}^*}$ is the same as $(M_{\mathfrak{p}})_g$ and the set $\theta_{\mathfrak{p}}((x_1,\ldots,x_{d-t}))$ is an independent set of general elements of the ideals $I_1(Q_{\mathfrak{p}})_g,\ldots,I_{d-t}(Q_{\mathfrak{p}})_g$, it follows that we can write the first factor as

$$e(Q_{\mathfrak{p}} \mid I'_U Q_{\mathfrak{p}} \mid M_{\mathfrak{p}})$$

Now we turn to the second factor $e(Q_g/\mathfrak{P} \mid \phi_{\mathfrak{p}}(x_{S-U}) \mid Q_g/\mathfrak{P})$.

First it will be convenient to take $\mathbf{x}_S$ in standard form. Next we observe that $Q_g/\mathfrak{P}$ is the ring $L(I_U,\mathfrak{p})$ which in this proof is abbreviated to L. Let $N = \mathrm{def}(\mathbf{x}_U)$. Then $\mathfrak{P} = \mathfrak{P}_N Q_g$, where $\mathfrak{P}_N = \mathfrak{P} \cap Q_N$ and hence, if $L_N = Q_N/\mathfrak{P}_N$, L is isomorphic to $(L_N)_g$. Next, as $\mathbf{x}_S$ is standard, $\mathbf{x}_{S-U}$ is an independent set of general elements of $I_{S-U}Q_N$, and so $\phi_{\mathfrak{p}}(\mathbf{x}_{S-U})$ is an independent set of general elements of the set of ideals $\mathbf{I}_{S-U}L_N$. Hence

$$\begin{aligned} e(Q_g/\mathfrak{P} \mid \phi_{\mathfrak{p}}(\mathbf{x}_{S-U}) \mid Q_g/\mathfrak{P}) &= e(L \mid \phi_{\mathfrak{p}}(\mathbf{x}_{S-U}) \mid L) \\ &= e((L_N)_g \mid \phi_{\mathfrak{p}}(\mathbf{x}_{S-U}) \mid (L_N)_g) \\ &= e(L_N \mid I_{S-U}L_N \mid L_N), \text{ by (12.1.9) and definition)} \\ &= e(L \mid I_{S-U}L \mid L), \text{ by (12.2.9).} \end{aligned}$$

We now state the above as a theorem. As it can be considered as a generalisation of (12.1.7), we number this Theorem 12.2.7.

THEOREM 12.2.7. *Let* $(Q,\mathfrak{m},k,d)$ *be a local ring,* $I_S = (I_1,\ldots,I_s)$ *be a large set of ideals of* Q, *and let* U *be a sub-set of* S = (1,...,s) *containing* u *integers. Let* I_U *denote the set of ideals* I_j *with* j *in* U, *and define* I_{S-U} *to be the set of ideals* I_j *with* $j \in S-U$. *Then, if* M *is any finitely generated* Q-*module,*

$$e(Q \mid I \mid M) = \Sigma_{\mathfrak{p}} e(Q_{\mathfrak{p}} \mid I'_U(\mathfrak{p})Q_{\mathfrak{p}} \mid M_{\mathfrak{p}})\ e(L(I_U,\mathfrak{p}) \mid I_{S-U}\, L(I_U,\mathfrak{p}) \mid L(I_U,\mathfrak{p})),$$

the sum being over all acceptable limit prime ideals $\mathfrak{p}$ *of* I_U, $L(I_U,\mathfrak{p})$ *being as defined in section* 11.3, *and* $I'_U(\mathfrak{p})$ *being the set of ideals* $I_j \in I_U$ *contained in* $\mathfrak{p}$.

3. The generalised degree formula.

Our purpose in this section is to develop a generalisation of the degree formula of chapter 9. In general terms this will be a formula for $e(I_1,\ldots,I_d \mid M)$ which takes the form

$$e(I_1,\ldots,I_d \mid M) = \Sigma a(I_1,\ldots,I_{d-1};M;v)v(I_d)$$

the sum being over all $\mathfrak{m}$-valuations v. However, the final form will contain a good deal of information about the coefficients $a(I_1,\ldots,I_{d-1};M;v)$.

We therefore consider an independent set of ideals $\mathbf{I} = (I_1,\ldots,I_{d-1})$ of a local ring $(Q,\mathfrak{m},k,d)$ and let $\mathbf{x} = (x_1,\ldots,x_{d-1})$ be an independent set of general elements of $\mathbf{I}$. Instead of I_d, we will introduce a variable ideal J such that $I_1,\ldots,I_{d-1},J$ is an independent set of ideals and we choose a general element y of J such that $x_1,\ldots,x_{d-1},y$ is $\mathfrak{m}$-primary.

In order to express our formula in a convenient form, we now introduce an ancillary function, the generalised degree function. Suppose that $(Q,\mathfrak{m},k,d)$ is a local *domain* and that $\mathbf{I}$, $\mathbf{x}$ are as above. We recall that, if $\mathfrak{P}$ is any minimal prime ideal over the ideal $x_1Q_g +\ldots+ x_{d-1}Q_g$ which meets Q in the zero ideal, then it is unique, (Theorem 11.31 ii) a)) and, by 11.21 iv), there exists such an ideal if and only if $X(\mathbf{I}) = (x_1Q_g +\ldots+ x_{d-1}Q_g)\cap Q = (0)$. Note that such a prime ideal $\mathfrak{P}$ is necessarily good. For, by 11.31 ii) a), $(Q_g)_{\mathfrak{P}}$ is regular of dimension d-1, whence $\dim Q_g/\mathfrak{P} \le 1$ and must be 1 since $\mathfrak{P} \neq \mathfrak{m}_g$.

DEFINITION. *Let* $(Q,\mathfrak{m},k,d)$ *a local domain and* $\mathbf{I}$, $\mathbf{x}$ *be as above. Then, if* $X(\mathbf{I}) = (0)$, *and* $\mathfrak{P}$ *is the unique good prime ideal minimal over* $x_1Q_g +\ldots+ x_{d-1}Q_g$ *meeting* Q *in* (0), *we define the generalised degree function* $d(Q \mid \mathbf{I},J)$ *to be* $e(L \mid JL \mid L)$, *where* $L = Q_g/\mathfrak{P}$ *and* J *ranges over all ideals of* Q *such that* $(\mathbf{I},J)$ *is an independent set of ideals. If* $X(\mathbf{I}) \neq (0)$, *then we define* $d(Q \mid \mathbf{I},J)$ *to be zero for all such* J.

First, note that the above definition is independent of the choice of $\mathbf{x}$. Secondly, suppose that y is a general element of J such that $\mathbf{x}$,y are independent general elements. Then as L is 1-dimensional, 11.22 implies that yL is a reduction of JL.

Hence we can define $d(Q \mid I,J)$ as $e(L \mid y \mid L)$.

Before stating our next lemma, it is convenient to introduce the following definitions.

DEFINITIONS. *Let* $\mathbf{p}$ *be a prime ideal of a local ring* Q *and let* $I = (I_1,\ldots,I_s)$ *be a set of ideals of* Q. *Then* $I'(\mathbf{p})$ *is defined to be the set of ideals* I_j *belonging to* I *which are contained in* $\mathbf{p}$ *and* $I''(\mathbf{p})$ *is the set of ideals* $I_j + \mathbf{p}/\mathbf{p}$ *of* $Q/\mathbf{p}$ *where* I_j *ranges over those* I_j *in* I *not in* $\mathbf{p}$.

LEMMA 12.3.1. *Let* $(Q,\mathfrak{m},k,d)$ *be a local ring,* $I = (I_1,\ldots,I_{d-1})$ *be an independent set of ideals of* Q, *and let* J *range over the set of ideals of* Q *such that the set* I, J *is independent. Let* M *be a finitely generated* Q-*module. Then*

$$e(Q \mid I,J \mid M) = \Sigma e(Q_{\mathbf{p}} \mid I'(\mathbf{p})Q_{\mathbf{p}} \mid M_{\mathbf{p}})d(Q/\mathbf{p} \mid (I''(\mathbf{p}), J(\mathbf{p}))$$

the sum being over all acceptable limit ideals $\mathbf{p}$ *of* I. $J(\mathbf{p})$ *denotes* $J + \mathbf{p}/\mathbf{p}$.

This is a special case of Theorem 12.2.7, with I_S taken to be the set (I, J) and I_U to be the set I. Since $S-U$ consists of the single ideal J and $L(I_U,\mathbf{p}) = L$ as defined above,

$$e(\, L(I_U,\mathbf{p}) \mid I_{S-U}\, L(I_U,\mathbf{p}) \mid L(I_U,\mathbf{p})) = e(L \mid JL \mid L) = d(Q/\mathbf{p} \mid I''(\mathbf{p}),J(\mathbf{p})).$$

LEMMA 12.3.2. *Let* $(Q,\mathfrak{m},k,d)$ *be a local domain, with field of fractions* F, *and let* $I = (I_1,\ldots,I_{d-1})$ *be an independent set of ideals of* Q. *Then the following two statements are true*

i) $d(Q \mid I,J\,) = \Sigma d(I,v)v(J)$

the sum being over all $\mathfrak{m}$-*valuations* v *on* F *satisfying the condition that* $\text{tr.deg}_k k_v = d-1$.

ii) *Further,* $d(I,v)$ *is, for each* v, *a function on the set of sets of independent ideals* $I = (I_1,\ldots,I_{d-1})$ *of* Q *which is symmetric and takes non-negative integral values.*

We suppose that $\mathbf{x} = (x_1,\ldots,x_{d-1})$ is a standard independent set of general elements of I, and that y is a general element of J such that $\mathbf{x}$, y is a standard independent set of general elements of I, J. $\mathbf{P}$ will denote the unique minimal prime ideal of $x_1Q_g +\ldots+ x_{d-1}Q_g$ meeting Q in (0).

By definition,

$$d(Q \mid I,J) = e(L \mid JL \mid L)$$

where $L = Q_g/\mathfrak{P}$ and so is a 1-dimensional local domain. Hence, by (12.2.8), noting that the residue field of L is k_g,

$$d(Q \mid I,J) = e(L \mid JL \mid L) = \Sigma\delta_i[k_i' : k_g]w_i(JL)),$$

summed for $i = 1,\dots,s'$, where $w_1,\dots,w_{s'}$ are $\mathfrak{m}'$-valuations on L, $\mathfrak{m}' = \mathfrak{m}_g/\mathfrak{P}$ being the maximal ideal of L, and k_i' is the residue field of w_i. Now consider the restriction w_i' of w_i to F (clearly $L \supseteq Q$). This will be an $\mathfrak{m}$-valuation on F which will take values multiples of a positive integer $\rho(i)$. Let $v_1,\dots,v_m$ be the distinct normalised $\mathfrak{m}$-valuations on F of which multiples are restrictions of the valuations w_i. Then we can write

$$d(Q \mid I,J) = \Sigma d(I,v_j)v_j(J)$$

summed over $j = 1,\dots,m$, where $d(I,v) = \Sigma\rho(i)\delta_i[k_i' : k]$, summed over those w_i whose restriction to F is v.

Next we have to prove that the residue fields k_i of v_i have transcendence degree d-1 over k. Choose N large enough to satisfy the following conditions. First, that some w_i that is an extension of v_i is a general extension of its restriction w_N to Q_N. Secondly, that $x_1,\dots,x_{d-1}$ belong to Q_N, so that $\mathfrak{P}$ is generated by elements of Q_N, i.e. $\mathfrak{P} = \mathfrak{P}_N Q_g$ where $\mathfrak{P}_N = \mathfrak{P} \cap Q_N$.

Now $L_N = Q_N/\mathfrak{P}_N$ is a 1-dimensional local domain. Hence if F_N is the field of fractions of L_N, the fact that $\mathfrak{P}$ is good implies, by Theorem 10.23, that

$$\text{tr.deg}_F F_N = N - \text{ht}\mathfrak{P} + \text{ht}(0) = N - (d-1) + 0 = N - d + 1.$$

Now the residue field k_N of L_N has transcendence degree N over k, and the residue field of w_N is algebraic over it. Hence the residue field of w_N has transcendence degree N over k. But, as F_N has transcendence degree $N - d + 1$ over F, the transcendence degree of the residue field of w_N over that of its restriction v_i to F is $N - d + 1$, whence k_i has transcendence degree $d - 1$ over k.

ii) If we now define $d(I,v)$ to be zero for v not one of $v_1,\ldots,v_s$, then we have defined the functions $d(I,v)$ for each v, since by the lemma of independence of $\mathfrak{m}$-valuations (Lemma 10. 31), the equation

$$d(Q \mid I,J) = \Sigma d(I,v)v(J)$$

uniquely determines the coefficients $d(I,v)$. The symmetry of $d(I,v)$ in the ideals of the set I, then follows from that of $d(Q \mid I,J)$. It is clear that $d(I,v)$ takes non-negative integer values.

The two lemmas above together constitute in essence most of the content of the generalised degree formula. All that really remains is to devise a notation in which to state it. First we require a definition.

DEFINITION. *Let* v *be an* $\mathfrak{m}$*-valuation on* $(Q,\mathfrak{m},k,d)$. *Let* $\mathbf{p}(v)$ *be the limit ideal of* v *(i.e., the set of elements* x *of* Q *such that* $v(x) = \infty$). *Then we say that* v *is good if*

$$\text{tr.deg.}_k k_v + \text{ht}\mathbf{p}(v) = d - 1.$$

(Note that this implies that $\mathbf{p}(v)$ is good, since $\text{tr.deg}_k k_v \leq \dim(Q/\mathbf{p}(v))-1$.)

We now introduce our changes of notation. Let v be an $\mathfrak{m}$-valuation on Q. First, the set of ideals $I_j \in I$ such that $v(I_j) = \infty$ is, in our present notation, denoted by $I'(\mathbf{p}(v))$. It seems worthwhile to denote this by, simply, $I'(v)$, since $\mathbf{p}(v)$ is determined by v. Similarly the set of ideals $I''(\mathbf{p}(v))$ will be denoted by $I''(v)$. Further, Q_v, M_v will mean $Q_{\mathbf{p}(v)}$, $M_{\mathbf{p}(v)}$. $Q(v)$ will denote $Q/\mathbf{p}(v)$, and, for any ideal J of Q, $J(v)$ will denote the ideal $(J + \mathbf{p}(v))/\mathbf{p}(v)$ of $Q(v)$. Finally, $d(v) = \dim Q(v)$, so that if v is good, the transcendence degree of k_v over k will be $d(v) - 1$. Then the generalised degree formula is the following theorem.

THEOREM 12.3.3. *Let* $(Q,\mathfrak{m},k,d)$ *be a local ring,* $I = I_1,\ldots,I_{d-1}$ *be an independent set of ideals of* Q *and let* J *range over the set of ideals of* Q *such that* I, J *is independent. Then, for any finitely generated* Q*-module* M,

$$e(Q \mid I, J \mid M) = \Sigma e(Q_v \mid I'(v)Q_v \mid M_v)d(I''(v),v)v(J(v))$$

the sum being over all good $\mathfrak{m}$*-valuations* v *on* Q, *for which* $\mathbf{p}(v)$ *is the intersection of a good minimal prime ideal* $\mathfrak{P}$ *of* $x_1Q_g + \ldots + x_{d-1}Q_g$ *with* Q, *where* $x_1,\ldots,x_{d-1}$ *is an independent set of general elements of* I.

We start with the formula

$$e(Q \mid I, J \mid M) = \Sigma e(Q_{\mathbf{p}} \mid I'(\mathbf{p})Q_{\mathbf{p}} \mid M_{\mathbf{p}})d(Q/\mathbf{p} \mid I''(\mathbf{p}),J(\mathbf{p}))$$

of Lemma 12.3.1, the sum being over all prime ideals $\mathbf{p}$ of Q of the form $\mathcal{P} \cap Q$, where $\mathcal{P}$ is a good minimal prime ideal of $x_1Q_g + \ldots + x_{d-1}Q_g$, for some set of independent general elements $(x_1,\ldots,x_{d-1})$ of I (note that the set of ideals $\mathbf{p}$ so defined is independent of the choice of $(x_1,\ldots,x_{d-1})$). We now substitute for the second factor of each term of the sum, using Lemma 12.3.2. The rest is a matter of making the necessary adjustments of notation.

To complete this section, we require some more information about the degree coefficients d(I,v).

THEOREM 12.34. *Let* $I' = (I_1,\ldots,I_{d-2})$ *be an independent set of general ideals of a local domain* Q *and let* K', K" *be ideals of* Q *such that* (I',K'), (I',K") *are both independent sets of general ideals of* Q. *Then, for each good* $\mathfrak{m}$*-valuation* v *on the field of fractions* F *of* Q,

$$d(I', K'K'',v) = d(I', K', v) + d(I', K'', v).$$

Let J range over all $\mathfrak{m}$-primary ideals of Q, so that I', K'K", J; I', K', J; I', K", J are all general sets of ideals. Then, by (12.2.2) and (12.2.4), we have

$$e(Q| I', K'K'', J |Q) - e(Q| I', K', J |Q) - e(Q| I', K'', J |Q) = 0$$

for all J. Now using Theorem 12.3.3, we can expand the left-hand side of the above equation as a linear combination of the valuations v(J), ranging over good $\mathfrak{m}$-valuations on Q. If v is such a valuation defined on the field of fractions of Q, then the coefficient of v(J) in this expansion is

$$d(I', K'K'', v) - d(I', K', v) - d(I', K'', v).$$

But by the lemma of independence of $\mathfrak{m}$-valuations (Lemma 10.31), this coefficient must be zero.

4. A final illustration.

In this final section we consider the case when the ideals $I_1,\ldots,I_{d-1},J$ are all assumed to be $\mathfrak{m}$-primary, and thereby link the development of this chapter to that of chapter 9. We also obtain a link with Teissier's work on mixed multiplicities.

We first consider the mixed multiplicity $e(Q \mid I \mid M)$, where we have written I_d in place of J, and, for the moment, we denote by I the set of ideals $I_1,\ldots,I_d$. Now let $N = (n_1,\ldots,n_d)$ be a set of non-negative integers. Then we will denote by I^N the product of ideals $I_1^{n_1}\ldots I_d^{n_d}$. Next suppose that K is a set of non-negative integers, $k_1,\ldots,k_d$, which satisfies the condition $k_1 + \ldots + k_d = d$. Then we will denote by $\binom{d}{K}$ the multinomial coefficient $(d!)/(k_1!)\ldots(k_d!)$. Next N^K will denote $n_1^{k_1}\ldots n_d^{k_d}$, and finally $e_Q(k_1,\ldots,k_d \mid M)$ will denote $e(Q \mid I(K) \mid M)$, where $I(K)$ consists of k_1 copies of I_1, k_2 of I_2, ..., k_d of I_d. Now consider the expression $e(Q \mid I^N,\ldots,I^N \mid M)$, which is defined since $I_1,\ldots,I_d$ are $\mathfrak{m}$-primary. Repeated use of Theorem 12.2.1 and symmetry and a simple combinatorial argument will yield the formula

$$e(Q \mid I^N,\ldots,I^N \mid M) = \Sigma\binom{d}{K}e_Q(k_1,\ldots,k_d \mid M)\, N^K$$

the sum being over all sets of non-negative integers $k_1,\ldots,k_d$ which satisfy $k_1 + \ldots + k_d = d$. In particular, $e(\,I \mid M)$ is the coefficient of $n_1 n_2 \ldots n_d$ in the above expression. This shows that the definition of mixed multiplicity used here agrees with that given by Bernard Teissier in Teissier [1973] Chapter 1, paragraph 2. In fact the definition given here is nothing more than an adaption of Teissier's presentation in that paper.

Now we turn to the degree formula, and we revert to our original notation in which I denoted a set of $d-1$ $\mathfrak{m}$-primary ideals $I_1,\ldots,I_{d-1}$ and J for the moment will be arbitrary. For simplicity we will assume that Q is a domain with field of fractions F.

A prime ideal $\mathfrak{p}$ of the form $\mathfrak{P}\cap Q$, where $\mathfrak{P}$ is a good minimal prime ideal of $x_1Q_g + \ldots + x_{d-1}Q_g$, must be the zero ideal of Q, since otherwise it must contain one of the ideals $I_1,\ldots,I_{d-1}$ which are all $\mathfrak{m}$-primary. It follows that

$$e(Q \mid I,J \mid M) = l(M\otimes_Q F)d(Q \mid I,J) = l(M\otimes_Q F)\Sigma d(I,v)v(J)$$

the sum being over all good $\mathfrak{m}$-valuations defined on F. Now we take J to be a principal ideal xQ of Q. Then we obtain

$$e(Q \mid I, xQ \mid M) = l(M \otimes_Q F)\Sigma d(I,v)v(x).$$

Next suppose we take the ideals $I_1,\ldots,I_{d-1}$ all equal to an ideal I. Then we have, except for a change in notation, obtained the result of Theorem 9.41 in the special case where $f = f_I$. The changes in notation are that $l(M \otimes_Q F)$ corresponds to $L_V(M)$, and $d(I,v)$ above is $\delta(v)d(f_I,v)$ in Theorem 9.41. It is with this link with chapter 9 that we conclude these notes.

BIBLIOGRAPHY

The following books are referred to in the text.

N.Bourbaki

[BAC] Elements de Mathematique: Algebre Commutative.
Chapters 1 to 7 published by Hermann, Paris, between 1961 and 1965.
Chapters 8 and 9 published by Masson, Paris, in 1984.

M.Herrmann, R.Schmidt and W.Vogel

[TNF] Theorie der Normal flachheit.
Teubner, Leipzig, 1976.

H.Matsumura

[M] Commutative Algebra, 2nd Edition.
Benjamin/Cummings Publishers, Reading, Mass., 1980.

M.Nagata

[LR] Local Rings.
Interscience Publishers, New York, 1960.

D.G.Northcott

[LRMM] Lessons on Rings, Modules and Multiplicities.
Cambridge University Press, 1968.

L.J.Ratliff Jr.

[CC] Chain Conjectures in Ring Theory.
Lecture notes in Mathematics no 647. Springer Verlag, Berlin, 1978.

O.Zariski and P.Samuel

[Z-S] Commutative Algebra, Vols. I and 2.
Van Nostrand, New York. Vol. I, 1958, Vol. 2, 1960.

In addition, the following book, closely related to these notes in subject matter, appeared after the lectures were given.

Stephen McAdam

Asymptotic Prime Divisors.
Lecture notes in Mathematics, Springer Verlag, Berlin, 1983.

The following papers are either mentioned in the text, where they are referred to by author and date, or were consulted in the preparation of the manuscript. This list does not constitute a complete bibliography of the asymptotic theory of ideals.

M.Auslander and D.A.Buchsbaum

[1958] Codimension and multiplicities
Annals Math. v68 (1958) pp. 625-657

[1959] Corrections to "Codimension and multiplicities"
Annals Math. v70 (1959) pp. 395-397

I.S.Cohen

[1946] On the structure and ideal theory of complete local rings.
Trans. Amer. Math. Soc. 59 (1946) pp. 54-106.

E.D.Davis and S.McAdam

[1977] Prime divisors and saturated chains.
Indiana U. Math. J. v26 (1977) pp. 653-662.

P.Eakin and A.Sathaye

[1976] Prestable ideals.
J. of Alg. 41 (1976) pp. 439-454.

K.H.Kiyek

[1981] Adwendung von Ideal-Transformationen.
Manuscripta Math. 34(1981) pp. 327-353.

J.R.Matijevic

[1976] Maximal ideal transformations of noetherian rings.
Proc. Amer. Math. Soc. 54 (1976) pp. 49-52.

Y.Mori

[1952] On the integral closure of an integral domain.
Mem. Coll. Sci. Univ. Kyoto. 27 (1952-3) pp. 249-256.
Errata. *ibid. 28.(1953-4). pp327-8.*

M.Nagata

[1955] On the derived normal ring of noetherian integral domains.
Mem. Coll. Sci. Univ. Kyoto. 29 (1955) pp. 293-303.

[1956] Note on a paper of Samuel concerning asymptotic properties of ideals.
Mem. Coll. Sci. Univ. Kyoto. 30 (1956-7) pp. 165-175.

D.G.Northcott and D.Rees

[1954a] Reductions of ideals in local rings.
Proc. Cam. Phil. Soc. 50 (1954) pp. 145-158.

[1954b] A note on reductions of ideals with an application to the generalised Hilbert function.
Proc. Cam. Phil. Soc. 50 (1954) pp. 353-359.

J.Querre

[1979] Sur la clôture integrale des anneaux noetheriens.
Bull. Soc. Math. (2) 103 (1979) pp. 71-76.

L.J.Ratliff Jr.

[1974] Locally quasi-unmixed noetherian rings and ideals of the principal class.
Pacific J.Math, 52 (1974) pp. 185-205.

D.Rees

[1955] Valuations associated with local rings (I).
Proc. London Math. Soc.(3) 5 (1955) pp. 107-128.

[1956a] Valuations associated with ideals (I).
Proc. London Math. Soc. (3) 6 (1956) pp.161-174.

[1956b] Valuations associated with ideals (II).
J. London Math. Soc. 31 (1956) pp. 221-8.

[1956c] Valuations associated with local rings (II).
J. London Math. Soc. 31 (1956) pp. 228-235.

[1961a] A note on analytically unramified local rings.
J. London Math. Soc. 36 (1961) pp. 24-8.

[1961b] Degree functions.
Proc. Cam. Phil. Soc. 57(1961) pp. 1-7.

[1961c] $\mathfrak{a}$-transforms of local rings and a theorem on multiplicities of ideals.
Proc. Cam. Phil. Soc. 57 (1961) pp. 8-17.

[1984] Generalisations of reductions and mixed multiplicities.
J. London Math. Soc. (2) 29 (1984) pp. 397-414.

[1986] The general extension of a local ring and mixed multiplicities.
Lecture notes in Mathematics no 1183. Springer Verlag. Berlin (1986) pp. 339-360.

P.Samuel

[1952] Some asymptotic properties of ideals.
Ann. Math. (2) 56(1952) pp. 11-21.

[1953] Commutative algebra (notes by D. Herzig).
Cornell University (1953).

B.Teissier

[1973] Cycles evanscents, sections planes, et conditions de Whitney.
Singularities à Cargese. Asterisque 7/8. (Soc. Math. de France. Paris (1973)).

INDEX

INDEX OF SYMBOLS